Natural Language Processing Projects

Build Next-Generation NLP
Applications Using AI Techniques

Akshay Kulkarni
Adarsha Shivananda
Anoosh Kulkarni

Apress®

Natural Language Processing Projects: Build Next-Generation NLP Applications Using AI Techniques

Akshay Kulkarni
Bangalore, Karnataka, India

Adarsha Shivananda
Bangalore, Karnataka, India

Anoosh Kulkarni
Bangalore, India

ISBN-13 (pbk): 978-1-4842-7385-2
https://doi.org/10.1007/978-1-4842-7386-9

ISBN-13 (electronic): 978-1-4842-7386-9

Managing Director, Apress Media LLC: Welmoed Spahr
Acquisitions Editor: Celestin Suresh John
Development Editor: Laura Berendson
Coordinating Editor: Aditee Mirashi
Copyeditor: Kimberly Burton Weisman

Cover designed by eStudioCalamar

Cover image by Pixabay (www.pixabay.com)

Distributed to the book trade worldwide by Apress Media, LLC, 1 New York Plaza, New York, NY 10004, U.S.A. Phone 1-800-SPRINGER, fax (201) 348-4505, e-mail orders-ny@springer-sbm.com, or visit www.springeronline.com. Apress Media, LLC is a California LLC and the sole member (owner) is Springer Science + Business Media Finance Inc (SSBM Finance Inc). SSBM Finance Inc is a **Delaware** corporation.

For information on translations, please e-mail booktranslations@springernature.com; for reprint, paperback, or audio rights, please e-mail bookpermissions@springernature.com.

Apress titles may be purchased in bulk for academic, corporate, or promotional use. eBook versions and licenses are also available for most titles. For more information, reference our Print and eBook Bulk Sales web page at http://www.apress.com/bulk-sales.

Any source code or other supplementary material referenced by the author in this book is available to readers on GitHub via the book's product page, located at www.apress.com/9781484273852. For more detailed information, please visit http://www.apress.com/source-code.

Printed on acid-free paper

To our families

Table of Contents

About the Authors

Akshay Kulkarni is a renowned AI and machine learning evangelist and thought leader. He has consulted several Fortune 500 and global enterprises on driving AI and data science–led strategic transformation. Akshay has rich experience in building and scaling AI and machine learning businesses and creating significant impact. He is currently a data science and AI manager at Publicis Sapient, where he is part of strategy and transformation interventions through AI. He manages high-priority growth initiatives around data science and works on various artificial intelligence engagements by applying state-of-the-art techniques to this space.

Akshay is also a Google Developers Expert in machine learning, a published author of books on NLP and deep learning, and a regular speaker at major AI and data science conferences.

In 2019, Akshay was named one of the top "40 under 40 data scientists" in India.

In his spare time, he enjoys reading, writing, coding, and mentoring aspiring data scientists. He lives in Bangalore, India, with his family.

Adarsha Shivananda is a lead data scientist at Indegene Inc.'s product and technology team, where he leads a group of analysts who enable predictive analytics and AI features to healthcare software products. These are mainly multichannel activities for pharma products and solving the real-time problems encountered by pharma sales reps. Adarsha aims to build a pool of exceptional data scientists within the organization to solve greater health care problems through brilliant training programs. He always wants to stay ahead of the curve.

His core expertise involves machine learning, deep learning, recommendation systems, and statistics. Adarsha has worked on various data science projects across multiple domains using different technologies and methodologies. Previously, he worked for Tredence Analytics and IQVIA.

He lives in Bangalore, India, and loves to read, ride, and teach data science.

Anoosh Kulkarni is a senior consultant focused on artificial intelligence (AI). He has worked with global clients across multiple domains and helped them solve business problems using machine learning (ML), natural language processing (NLP), and deep learning. Currently, he is working with Subex AI Labs. Previously, he was a data scientist at one of the leading e-commerce companies in the UAE. Anoosh is passionate about guiding and mentoring people in their data science journeys. He leads data science/machine learning meetups in Bangalore and helps aspiring data scientists navigate their careers. He also conducts ML/AI workshops at universities and is actively involved in conducting webinars, talks, and sessions on AI and data science. He lives in Bangalore with his family.

About the Technical Reviewer

Aakash Kag is a data scientist at AlixPartners and the co-founder of the Emeelan and EkSamaj application. He has six years of experience in big data analytics. He has a postgraduate degree in computer science with a specialization in big data analytics. Aakash is passionate about developing social platforms, machine learning, and the meetups where he often talks.

Acknowledgments

We are grateful to our families for their motivation and constant support.

We want to express our gratitude to out mentors and friends for their input, inspiration, and support. A big thanks to the Apress team for their constant support and help.

Finally, we would like to thank you, the reader, for showing an interest in this book and making your natural language processing journey more exciting.

Note that the views and opinions expressed in this book are those of the authors.

CHAPTER 1

Natural Language Processing and Artificial Intelligence Overview

In recent years, we have heard a lot about artificial intelligence, machine learning, deep learning, and natural language processing. What are they? Are they all the same? How do we differentiate between them?

In 1956, an American computer scientist named John McCarthy coined the term *artificial intelligence*, a subfield of computer science. Artificial intelligence (AI) is a machine's ability to think and learn. The concept of AI is to make machines capable of thinking and learning like the human brain.

Before getting into more detail, it is important to clarify one thing about AI: most of the population falsely considers AI a technology, whereas it is a concept in which machines can deal with tasks intelligently.

Today, it is an umbrella that comprehends everything from data interpretation to humanoid robotics. It has gained prominence recently due to an increase in computation power and huge data. It can identify patterns in data more efficiently than humans, enabling any business to acquire more insight from their own data.

Is AI the same as machine learning? Not really. The two terms can often be used interchangeably, but they are not the same. Artificial intelligence is a broader concept, while machine learning is one part of the artificial intelligence paradigm and can be considered a subfield of AI. Under the AI paradigm, there are several concepts, like machine learning, natural language processing, deep learning, and many more.

You can understand machine learning as a portion of AI focused on making machines learn from data and enable machines to think and make decisions without

© Akshay Kulkarni, Adarsha Shivananda and Anoosh Kulkarni 2022
A. Kulkarni et al., *Natural Language Processing Projects*, https://doi.org/10.1007/978-1-4842-7386-9_1

human assistance or intervention. The entire machine learning concept assumes that we should give machines access to data and learn from it themselves.

Similarly, deep learning mainly deals with unstructured data like images, text, and audio.

And when specifically dealing with text data, we leverage natural language processing. Natural language processing (NLP) needs no introduction in the present scenario. It's one of the fields where study and research are very active, and the interest has multiplied in the last few years. Even though the fundamentals of NLP are quite easy, the advanced topics are complicated to understand.

That's where machine learning, deep learning, and other state-of-the-art algorithms become so pivotal. Figure 1-1 is a Venn diagram of the AI paradigm.

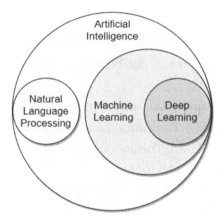

Figure 1-1. *AI paradigm*

Before getting into projects, let's understand some of these concepts and techniques required to solve upcoming projects at a very high level. An in-depth understanding of these concepts is beyond the scope of this book.

Machine Learning

Machine learning can be defined as the machine's capability to learn from experience(data) and make meaningful predictions without being explicitly programmed.

It is a subfield of AI which deals with building systems that can learn from data. The objective is to make computers learn on their own without any intervention from humans.

There are three main categories in machine learning.

- Supervised learning
- Unsupervised learning
- Reinforcement learning

Supervised Learning

Supervised learning is a part of machine learning where labeled training data is leveraged to derive the pattern or function and make models or machine learning. Data consists of a dependent variable (target label) and the independent variables or predictors. The machine tries to learn the function from labeled data and predict the output on unseen data.

Unsupervised Learning

Here the machine learns the hidden pattern without leveraging labeled data, so training doesn't happen. Instead, these algorithms learn based on similarities or distances between features to capture the patterns.

Reinforcement Learning

Reinforcement learning is a process of maximizing the reward by taking action. These are goal-oriented algorithms that learn how to reach a goal through experience.

Going forward, we discuss all the algorithms which are part of these categories briefly. You also learn how natural language processing works and how deep learning is solving NLP problems.

Figure 1-2 explains all the categories and subcategories.

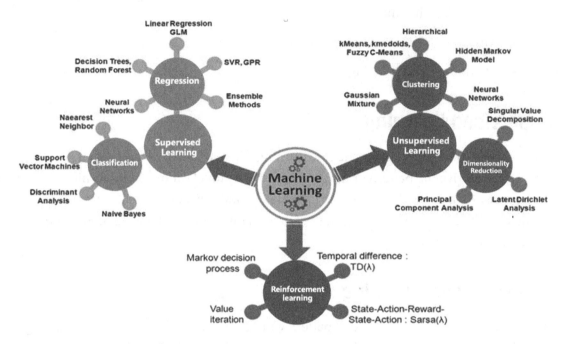

Figure 1-2. *Machine learning categories*

Supervised Learning

Learning from labeled data is called *supervised learning,* of which there are two types: *regression* and *classification.*

Regression

Regression is a statistical predictive modeling technique that finds the relationship between the dependent variable and independent variables. Regression is used when the dependent variable is continuous; prediction can take any numerical value.

The following are the few regression algorithms widely used in the industry. Let's get into the theory to better understand these algorithms.

- Linear regression

- Decision tree

- Random forest

- Support-vector machines
- GBM
- XGBOOST
- ADABOOST
- LightGBM

Classification

Classification is a supervised machine learning technique where the dependent or output variable is categorical. For example, spam/ham, churn/not churned, and so on.

- In *binary classification*, it's either yes or no. There is no third option. For example, the customer can churn or not churn from a given business.
- In *multiclass classification*, the labeled variable can be multiclass, for example, product categorization of an e-commerce website.

The following are the few classification algorithms widely used in the industry.

- Logistic regression
- Decision tree
- Random forest
- Support-vector machine
- GBM
- XGBoost
- AdaBoost
- LightGBM

Let's briefly look at each of these algorithms.

Linear Regression

Regression is the most frequently used term in analytics and data science. It captures the relationship between the target feature and independent features. And it is used when you want to predict a continuous value. Figure 1-3 demonstrates how linear regression works.

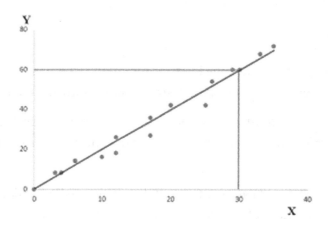

Figure 1-3. *Linear regression*

In Figure 1-3, points are plotted using the x-axis and the y-axis. The aim is to find the relation between *x* and *y* by drawing an optimal line close to all the points on the plane (minimizing error). The slope of the line derives from the following equation.

$$Y = A_0 + A_1X$$

Y = dependent variable
X = independent variable
A_0 = intercept
A_1 = coefficient of X
For example,

$$Age = Ht * 4 + wt * 3 + 2$$

Here we are forming a relationship between height and weight with age. Given the height and weight, you can calculate age.

There are few assumptions in linear regression.

- There should always be a linear relationship between dependent variables and independent variables.

- Data should be normally distributed.

- There should not be any collinearity between independent variables. Multicollinearity refers to a strong linear relationship between independent variables. These correlated variables act as redundant and need to be treated.

- There is *homoscedasticity*, which means the variance of errors should be constant over time. Variance shouldn't be higher for higher output values and lower for lower output values.

- The error terms should not have any correlation among themselves.

Once you build the model, you want to know how well the model is performing. To do this, you can use metrics.

- *R-squared* (R^2) is the most widely used metric to calculate the accuracy of linear regression. The R^2 value shows how the variance of dependent variables is explained through independent variables. R-squared ranges from 0 to 1.

- The *root-mean-square error* (RMSE) shows the measure of the dispersion of predicted values from actual values.

Logistic Regression

We discussed how to predict a numerical value using linear regression. But we also encounter classification problems where dependent variables are binary classes like yes or no, 1 or 0, true or false, and so on. In that case, you need to use *logistic regression,* which is a classification algorithm. Logistic regression is an extended version of linear regression, but we use a log of odds to restrict the dependent variable between zero and one. The following is the logic function.

$$\log\left(\frac{p}{1-p}\right) = \beta_0 + \beta(Age)$$

(p/1 – p) is the odds ratio. B0 is constant. B is the coefficient.

Figure 1-4 shows how logistic regression works.

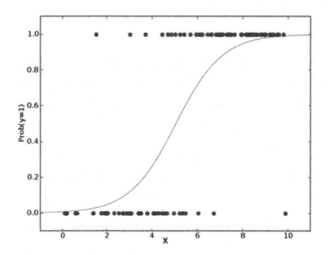

Figure 1-4. *Logistic regression*

Now let's look at how to evaluate a classification model.

- **Accuracy** is the number of correct predictions divided by the total number of predictions. The values lie between 0 to 1, and to convert it into a percentage, multiply the answer by 100. But only considering accuracy as the evaluation parameter is not an ideal thing to do. For example, you can obtain very high accuracy if the data is imbalanced.

- A **confusion matrix** is a crosstab between actual vs. predicted classes. You can use it for binary and multiclass classification.

    Figure 1-5 represents a confusion matrix.

		Predicted Class	
		Yes	No
Actual Class	Yes	TP	FN
	No	FP	TN

Figure 1-5. *Confusion matrix*

- **ROC curve**: A *receiver operating characteristic* (ROC) curve is an evaluation metric for classification tasks. A ROC curve plot has a false positive rate on the x-axis and a true positive rate on the y-axis. It says how strongly the classes are distinguishing when the thresholds are varied. Higher the value of the area under the ROC curve, the higher the predictive power. Figure 1-6 shows the ROC curve.

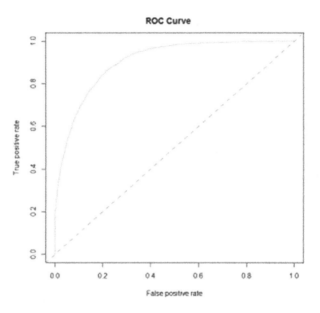

Figure 1-6. ROC curve

Linear and logistic regression are traditional ways of doing things that use statistics to predict the dependent variable. But there are few drawbacks to these algorithms. The following describes a few of them.

- Statistical modeling must satisfy the assumptions discussed previously. If they are not satisfied, models won't be reliable and thorough fit random predictions.

- These algorithms face challenges when data and target feature is nonlinear. Complex patterns are hard to decode.

- Data should be clean (missing values and outliers should be treated).

There are other advanced machine learning concepts like a decision tree, random forest, SVM, and neural networks to overcome these limitations.

The decision is a type of supervised learning in which the data is split into similar groups based on the most important variable to the least. It looks like a tree-shaped structure when all the variables split hence the name tree-based models.

The tree comprises a root node, decision node, and leaf node. A decision node can have two or more branches, and a leaf node represents a decision. Decision trees are capable of handling any type of data, whether it is quantitative or qualitative. Figure 1-7 shows how the decision tree works.

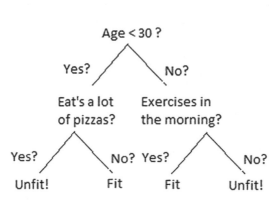

Figure 1-7. *Decision tree*

Let's explore how tree splitting happens, which is the key concept in decision trees.

The core of the decision tree algorithm is the process of splitting the tree. It uses different algorithms to split the node and different for classification and regression problems. Now, let's try to understand it.

Classification

- The **Gini index** is a probabilistic way of splitting the trees. It uses the sum of the square of probability for success and failure and decides the purity of the nodes.

- **CART** (classification and regression tree) uses the Gini method to create splits.

- A **Chi-square** is a statistical significance between subnodes, and the parent node decides the splitting. Chi-square = ((Actual – Expected)^2 / Expected)^1/2.

- **CHAID** (Chi-square Automatic Interaction Detector) is an example of this.

Regression

- **Variance reduction** works to split the tree based on the variance between two features (target and independent feature).

- **Overfitting** occurs when the algorithms tightly fit the given training data but inaccurately predict the untrained or test data outcomes.

This is the case with decision trees as well. It occurs when the tree is created to perfectly fit all samples present in the training data set, affecting the accuracy of test data.

Random Forest

Random forest is the most widely used machine learning algorithm because of its flexibility and ability to overcome the overfitting problem. Random forest is an ensemble algorithm that is an ensemble of multiple decision trees—the higher the number of trees, the better the accuracy.

The random forest can perform both classification and regression tasks. The following are its advantages.

- Random forest is insensitive to missing values and outliers.

- It prevents the algorithm from overfitting.

How does it work? It works on bagging and bootstrap sample techniques.

- Randomly takes the square root of m features and a 2/3 bootstrap data sample with a replacement for training each decision tree randomly and predicts the outcome.

- An n number of trees are built until the out-of-bag error rate is minimized and stabilized.

- Compute the votes for each predicted target and consider the mode as a final prediction in terms of classification.

Figure 1-8 shows the working of the random forest model.

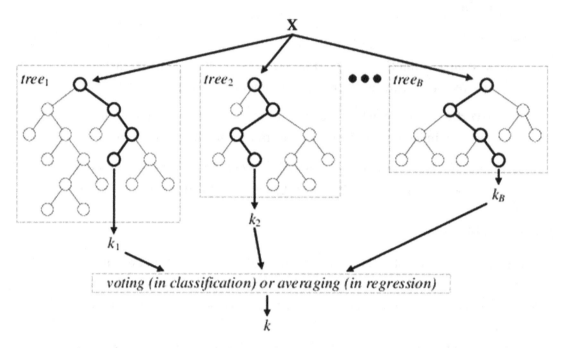

Figure 1-8. *Random forest*

Support-Vector Machines

SVM tries to find the best hyperplane which separates the data into groups based on the target feature. Figure 1-9 represents the working of SVM.

Figure 1-9. *SVM*

Consider the graph Figure 1-9 as two groups: circles and boxes. Since there are only two features, separating these groups becomes easier, and using a line to do so. But as the number of features increases, dimensionality also increases and uses a hyperplane or a set of hyperplanes to separate the classes. These planes are called a *decision boundary*. Anything that falls on one side of the line is classified as a *circle*, and the other side is a *box.*

How to decide the best hyperplane? Figure 1-10 demonstrates the hyperplane concept.

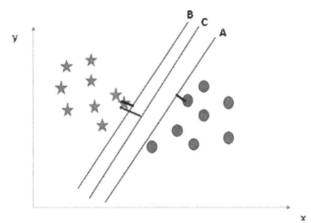

Figure 1-10. *Hyperplane*

Margin width is the distance between the plane and the nearest data point. SVM tries to maximize the margin width to achieve the goal of choosing the right hyperplane. As shown in the image, line C is the best hyperplane as the margin is high.

Nonlinear Data: The Kernel Trick

The data distribution is not always linear. Most of the time, the data distribution is nonlinear, and linear hyperplanes fail to classify the data. To solve this problem, SVM introduced the *kernel trick* (see Figure 1-11), which projects data multidimensionally to perform learning tasks.

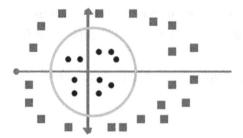

Figure 1-11. *Kernel trick*

Neural Networks

Neural networks are machine learning algorithms inspired by the human nervous system, which processes information and takes actions.

Millions of neurons are interconnected, and each neuron learns or finds the pattern based on the training data. The network tries to find the hidden rules automatically without human intervention. These algorithms outperformed all other machine learning techniques, especially on unstructured data like image and text. Figure 1-12 shows the architecture of the neural network.

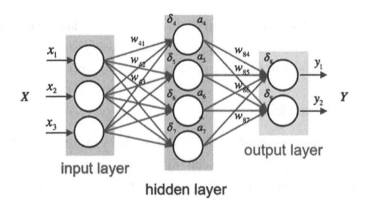

Figure 1-12. *Neural network*

You need to find the weights and biases that are predicted. The training of neural networks is done by fine-tuning the weights and biases through multiple iterations.

14

- Feedforward: Taking the input through the input layer, passing it on to the hidden layer where the weights and bias are added randomly for the first iteration, and finally predicting the output as the sum product of weights and feature value plus bias.

- Backpropagation: Computes error between the predicted and actual output from the previous step. This error is propagated backward, and weights and biases are updated accordingly using optimizers to minimize the loss function.

One cycle of feedforward and backpropagation is called an *epoch*. And such multiple epochs are carried out to improve the accuracy.

The *activation function* introduces non-linearity to map the input to the output accordingly. Multiple activation functions are based on the target feature, such as Sigmoid, which is mostly used in binary classification; similarly, Softmax is predominantly used in multiclass classification.

The following are key terms or components in deep learning.

- Weights

- Bias

- Loss function

- Optimizers: gradient descent, Adam, AdaDelta, RMS Prop

- Activation functions: sigmoid, softmax, ReLU

Deep Neural Networks

Characteristics

- The number of hidden layers is more than one

- The previous layer's output is fed into the next layers

- Each layer extracts features

- Nonlinear functions capture complex hidden features

Figure 1-13 shows a flow diagram.

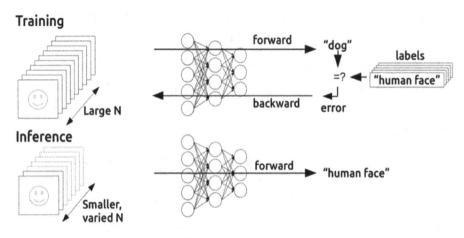

Figure 1-13. *Deep neural network*

Convolutional Neural Networks

Convolution neural networks are a type of deep neural network with a minor change in architecture. These are specifically designed for processing images. CNN has a couple of other layers, which makes the implementation more efficient. CNN takes every image pixel as a feature with a 3D array (for color images) and processes it to classify it.

There are multiple custom layers in CNN, including the following.

- Convolution layer

- Max pooling layer

- Fully connected layer

- Dropout layer

The *convolution layer* acts as a filter that extracts the feature present in an image. The values of the filters are derived in the training process. Figure 1-14 shows the workings of CNN.

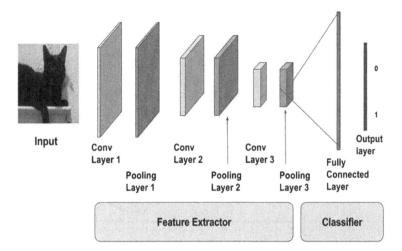

Figure 1-14. *CNN*

RNN

Just as CNNs are specifically designed for images, there are recurrent neural networks to process sequence-based data like text. The idea is that text data have a certain sequence attached to it. The data itself is sequential, and that adds meaning to the sentence. RNNs also capture the sequential nature of data.

RNN works in loops. The data flows in circles and thereby learning not only current state input but previous input as well. The concept of backpropagation through time comes into the picture; it is similar to the backpropagation of a normal neural network, but it goes back in time and updates the weights.

But the biggest challenge in RNN is vanishing gradients which means when the gradients are too small, the learning of the model will stop. Models like *long short-term memory* (LSTM) networks and *gated recurrent units* (GRUs) overcome problems using the gates mechanism.

Unsupervised Learning

Unsupervised learning is a machine learning category where the labeled data is not leveraged, but still, inferences are discovered using the data at hand. You need to find the patterns without the dependent variables to solve the business problems. Figure 1-15 shows the clustering outcome.

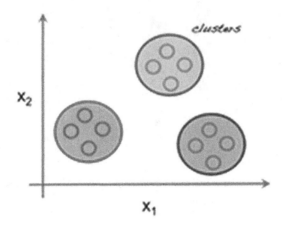

Figure 1-15. *Clustering*

There are two main types of unsupervised learning techniques: *clustering* and *dimensionality reduction*.

Clustering

Grouping similar things into segments is called *clustering*. Similar "things" are not only data points but a collection of observations that are

- Similar to each other in the same group

- Dissimilar to the observations in other groups

There are mainly two important algorithms that are highly being used in the industry. Let's briefly look at how algorithms work before getting into projects.

k-means Clustering

k-means is the efficient and widely used clustering technique that groups the data based on the distance between the points. The objective of k-means clustering is to minimize total variance within the cluster, as shown in Figure 1-16.

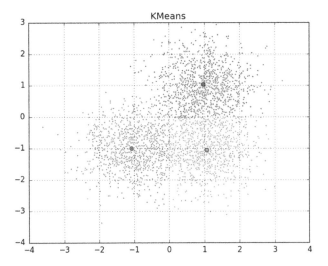

Figure 1-16. *k-means*

The following steps generate clusters.

1. Use the elbow method to identify the optimum number of clusters. This act as *k*.

2. Select random *k* points as cluster centers from the overall observations or points.

3. Calculate the distance between these centers and other points in the data and assign to the closest center cluster that point belongs to using any distance metrics.

   - Euclidean distance

   - Manhattan distance

   - Cosine distance

   - Hamming distance

4. Recalculate the cluster center or centroid for each cluster.

5. Repeat steps 2, 3, and 4 until the same points are assigned to each cluster, and the cluster centroid is stabilized.

Hierarchical Clustering

Hierarchical clustering is another type of clustering technique that also uses distance to create groups. The following steps generate clusters.

1. Hierarchical clustering starts by creating each observation or point as a single cluster.

2. It identifies the two observations or points that are closest together based on distance metrics.

3. Combine these two most similar points and form one cluster.

4. This continues until all the clusters are merged and form one final single cluster.

5. Finally, using a dendrogram, decide the ideal number of clusters.

You cut the tree to decide the number of clusters. The cutting of the tree is one where there is a maximum jump of one level to another, as shown in Figure 1-17.

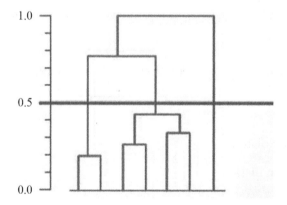

Figure 1-17. *Hierarchical clustering*

Usually, the distance between two clusters has been computed based on Euclidean distance. Many other distance metrics can be leveraged to do the same.

Dimensionality Reduction

Dimensionality reduction is the process of decreasing the number of features from an entire data to a few significant principal features to increase the accuracy of our modeling practices and reduce computational challenges. For example, if we are working on predictive modeling and the number of variables is more than 1000, there is a good possibility that our algorithms might fail to perform. Using any dimensionality reduction techniques, you can bring the number of features while capturing most of the context/information from all the features.

- It helps when independent variables have a correlation between them, which is called *multicollinearity*. We might have encountered this more often while working on our machine learning algorithms. You can also use *principal component analysis* (PCA) and *linear discriminant analysis* (LDA) to solve such a challenge.

Reinforcement Learning

You saw two well-known machine learning categories, supervised and unsupervised learning. In recent days, reinforcement learning is also gearing up to solve a lot of business problems. There is a lot of research in this field, and it has already showcased its potential. We all know that machines playing games against humans are already a talking point about how machines are winning the contest. AlphaGo is the best example.

How does it work?

Reinforcement learning is learning from experience and mistakes. There is a goal to achieve in every setting, and for every positive move, a reward is given. In the same way, for every negative move, it penalizes the machine. It learns what kind of actions it should take to gain more rewards and reach the goal. It is a continuous sequential process, and, in some cases, there might be no information at the beginning of the process.

It has five components to carry out the process, as shown in Figure 1-18.

- Environment

- State

- Agent

- Action

- Reward or penalty

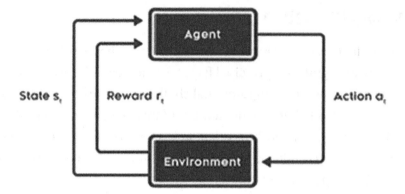

Figure 1-18. *Reinforcement learning*

1. The process starts from state S0, which is the environment.

2. The agent takes action A0 after this.

3. The state has changed from S0 to S1.

4. Reward R1 is provided to the agent.

5. This RL loop repeats until the goal is achieved.

NLP Concepts

Unstructured Data

When data is stored in a structured manner with columns and rows, it's called *structured data*. But more than 80% of the data cannot be used in the same way because of its unstructured behavior. But there is infinite insight hidden within that data as well. Unstructured data can be text, image, video, and so on.

Natural Language Processing

Collecting, understanding, processing, and generating insights out of text data is called *natural language processing*.

Text data is available everywhere. The exponential increase in social media, retail product reviews, entertainment reviews, and healthcare documents generates an unimaginable amount of text data. Leveraging these kinds of data is critical in the era of AI coupled with business.

Text Preprocessing

Text data is so unstructured and clumsy. There is a lot of junk present in the text, and it requires mandatory preprocessing to be done on top of raw data. It's vital because unwanted text can change the direction of the results it produces. So, let's quickly look at different data preprocessing techniques to clean the text data.

- Lowercasing

- Removal of punctuations and stop words

- Spelling correction

- Text standardization

- Text normalization

    - Stemming

    - Lemmatization

We understand concepts related to these techniques in this chapter and implement them in the rest of the book as we move along.

- **Lowercasing** converts text to lowercase to avoid duplication of features.

- **Removal of punctuations and stop words**: Punctuation doesn't add much value to a sentence's meaning. In the same context, stop words (*the, a, it, is*, and so on) are meaningless depending on the use cases. Their presence across the corpus skew the results. If we remove these words, weight distribution is more accurate, and crucial words get more importance.

- Spelling correction: The raw data we encounter has a lot of spelling mistakes, and using it without correcting creates redundancy. You need to correct it to some extent and start using it for better results using certain packages.

- **Text standardization**: Abbreviations are another biggest challenge in text data. Whenever we are dealing with reviews and comments, there is high chance customers are using the abbreviation. You need to take care of this; this process is equally important.

- **Text normalization**: The words *clean, cleaning,* and *cleaned* mean the same but are used in different tenses. Normalizing these words is called *text normalization.*

  - **Stemming**: Normalizing texts to the base word is called *stemming.* For example, *clean, cleaning,* and *cleaned* are stemmed to *clean.*

  - **Lemmatization**: Lemmatization gets to the root word, considering the tenses and plurality of a word. For example, *best* and *good* can mean the same thing. After lemmatization, the result for both words is *good.*

Text to Features

Since machines or algorithms cannot understand the text, a key task in NLP is to convert text data into numerical data called *features.* There are different techniques to do this. Let's discuss them briefly.

One-Hot Encoding (OHE)

OHE is the basic and simple way of converting text to numbers or features. First, it converts all the tokens present in the corpus into columns, as shown in Table 1-1. After that, against every observation, it tags a 1 if the word is present; otherwise, it tags a 0.

Table 1-1 demonstrates OHE.

Table 1-1. *OHE*

	One	Hot	Encoding
One-Hot	1	1	0
Hot	0	1	0
Encoding	0	0	1

Count Vectorizer

The drawback to this approach is that if the word appears multiple times in a sentence, it has the same importance as any other word that has appeared only once. To overcome this, there is the count vectorizer where everything else remains the same but counts the tokens present in that observation instead of tagging it as 1 or 0.

Table 1-2 demonstrates a count vectorizer.

Table 1-2. *Count Vectorizer*

	AI	**new**	**Learn**	**.......**
AI is new AI is everywhere	2	1	0	....
Learn AI Learn NLP	1	1	2	....
NLP is cool	0	0	1	....

Term Frequency–Inverse Document Frequency (TF-IDF)

Even a count vectorizer cannot answer all questions. If the length of sentences is inconsistent and a word is repeating in all the sentences, then it becomes tricky. TF-IDF addresses these problems.

- **Term frequency** (TF) is defined as the "number of times the token appeared in a corpus doc divided by the total number of tokens."

- **Inverse document frequency** (IDF) is the total number of corpus docs in the overall docs divided by the number of overall docs with the selected word. This provides more weight to rare words in the corpus.

Multiplying these two gives the TF-IDF vector for a word in the corpus.

$$tfidf_{i,j} = tf_{i,j} \cdot idf_i$$

$$tf_{i,j} = \frac{Number\ of\ times\ term\ i\ appears\ in\ document\ j}{Total\ number\ of\ terms\ in\ document\ j}$$

$$idf_i = log\left(\frac{Total\ number\ of\ documents}{Number\ of\ documents\ with\ term\ i\ in\ it}\right)$$

Word Embeddings

Even though TF-IDF is most widely used, it doesn't capture the context of a word or sentence. Word embeddings solve this problem. Word embeddings capture context and semantic and syntactic similarities between words.

Using shallow neural networks, we generate a vector that captures the context and semantics.

In recent years, there has been a lot of advancements in this field, particularly by

- word2vec

- GloVe

- fastText

- ELMo

- Sentence-BERT

- GPT

For more information about these techniques, please refer to our book *Natural Language Processing Recipes: Unlocking Text Data with Machine Learning and Deep Learning Using Python* (Apress, 2021).

Applications

Text Classification

This is where your machine learning starts interacting with text data. The classic example is your spam classification. Independent variables are the feature extracted using the steps discussed (text to features) earlier. Dependent variables depend on whether the mail is spam or ham. Using a machine learning classification algorithm, let's build a model that predicts whether mail is spam or ham. Many deep learning techniques like RNN and LSTM also obtain better accuracy.

In the coming chapters, we build both binary and multiclass text classification models to solve different business use cases.

Text Summarization

The other powerful application of NLP coupled with deep learning is text summarization. It is the process of creating short notes out of a long and detailed document. There is an abundant amount of text data, and reading every word takes a lot of time. That's where the industry saw its need and worked toward building state-of-the-art text summarization solutions using deep learning and reinforcement learning.

You learn how to implement text summarization using deep learning in one of the projects in this book.

Text Generation

The AI world is moving from understanding to generation. Be it image or text, we are witnessing a good amount of traction. For instance, an AI-generated painting sold for millions of dollars.

Text generation uses neural language models at its core. Recurrent neural networks are the foundation of neural language models, with variants LSTM and GRU networks. The latest work in encoder-decoder coupled with attention mechanism approaches achieved good results in this field.

Let's further discuss this concept with the application using RNNs and transformers to build a text generator.

And there are many more such applications.

- **Text regression**: "Predicting time to resolve" an issue or complaint by looking at the description of the complaint

- **Text similarity**: Search engines, content-based recommendation systems, resume shortlisting based on the job description, and many others

- **Entities extraction**: Extracting important and required words or a set of words from the document

- **Conversational AI**: Chatbots and Q&A systems

Now that you understand the fundamentals of machine learning, deep learning, and NLP, let's look at the life cycle (step by step) leveraging the concepts to solve business problems.

The AI Life Cycle

There are ten critical steps everyone should follow to solve any AI and data science problems. Solving a problem using data science is not an easy task. It's multidisciplinary and includes different steps, right from business understanding to productionizing for real-time output. To simplify this long process, let's divide the complete data science and AI life cycle into ten different steps.

1. Understanding and defining the business problem

2. Translating the business problem into a machine learning problem and creating a solution approach/architecture design

3. Data collection and understanding

4. Data cleaning and preprocessing

5. Exploratory data analysis

6. Feature engineering and selection

7. Model building, tuning, and selection

8. Model testing and validation

9. Prediction and insights

10. Deployment or productionization and model maintenance

Let's dive into each of these steps.

Understanding and Defining the Business Problem

The entire data science process stands on this step which is identifying/defining the problem. Just because data science is thrilling doesn't mean you can randomly create a problem statement.

Understanding the business, problem, bottlenecks, pain points, and business goals is very important. Data science is not about making your product or company cool; rather, it's about solving the actual problems that a business is facing to make a significant impact on increasing revenue, reducing cost, or customer experience/satisfaction.

Once the problem is defined well, let's say to decrease the business's churn rate, then start exploring the possible solutions, data requirements, and so on.

Translating the Business Problem into an ML Problem and Creating a Solution Approach/Architecture Design

The critical part of the data science process is converting a business problem into an analytical or machine learning approach. Few problems seem straightforward, but not all.

It takes mastering all the algorithms, business knowledge, and experiences to create a solid architecture to solve the problem effectively and efficiently.

1. Validate if a problem can be solved by simple analytical methods like exploratory data analysis, bivariate analysis, and correlation. You can get most of the answers through these steps.

2. Next, go deeper into the problem to see whether it's supervised, unsupervised, or reinforcement learning.

3. If it's a supervised learning problem, does it belong to regression or classification? Based on all of these, use machine learning or statistical algorithms to solve the problem. Figure 1-19 shows the scenarios.

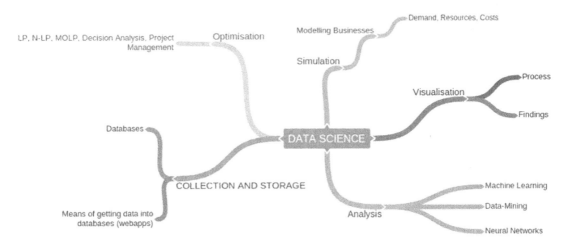

Figure 1-19. *Scenarios*

Data Collection and Understanding

This part involves thinking through what data is required to solve the given problem and finding ways to get that data. Data can be

- Internal databases like transaction data

- Freely available open source data like population

- Purchasing external data

Data can also be in any format.

- Structured data (data present in table format)

- Unstructured data (image, text, audio, and video)

Remember the expression, "garbage in, garbage out." Hence, data collection is the most critical step.

Data Cleaning and Preprocessing

Now that the raw data is ready, you need to process it before starting any type of analysis. Many times, data can be quite messy, and it would corrupt the analysis. The most time-consuming step in the entire data science process is data cleaning and preparation.

- Data cleaning

  - Missing value treatment

  - Imputation methods

  - Detect/remove outliers

  - Eliminating duplicate rows

- Combining multiple data sources

- Transforming data

Exploratory Data Analysis

A more interesting step in the overall process is where we torture the data to find hidden patterns, insights and visualize to tell a story from the data.

The following are the most common things done as part of EDA.

- Hypothesis testing

- Descriptive statistics or univariate analysis

- Inferential statistics or bivariate analysis

- Pivots, plots, visuals, and so on

Feature Engineering and Selection

Feature engineering is a process where that leverages domain and business knowledge to create more features or variables that has meaningful relationships with the target feature. Proper feature engineering increases the accuracy of the machine learning models and provides valuable and relevant insights. It is an art.

Feature extraction extracts new features from raw data and creates valuable variables that stabilize the machine learning model. They are manually created with spending a lot of time on actual data and thinking about the underlying problem, structures in the data, and how best to expose them to predictive modeling algorithms. It can be a process of aggregating or combining variables to create new ones or splitting features to create new features.

Not all features are created to affect the outcome of the model. There are only a few that are useful, and most of them are junk. These junk variables need to be removed to achieve better results. Regularization methods like LASSO/ridge regression and Stepwise regression are examples that automatically perform feature selection.

The feature engineering process

- Derives features and KPIs from the existing ones, which adds more meaning and relation to the target feature

- Creates more sensible features from the business context

- Validates the features using it in the model

- Repeats the process until you find the best features for the model

The following are some examples.

- Extracting day, time, and weekday from dates

- Extracting the region from a location

- Creating grouped data from numerical features

Model Building, Tuning, and Selection

With all the data in place and processed, split the data into training and testing data and get into the model building phase. Choose the appropriate algorithm to solve the problem. It's not a one-time process; it's an iterative process. You must try multiple algorithms before selecting the final one.

There are so many algorithms and tuning parameters to get the highest possible accuracy. But manually performing that task takes a lot of time. There are many model-tuning libraries like grid search that provide the best model to use for implementation.

Model Testing and Validation

It is very important to test the model on unseen data and see if it's not overfitting. If the model performs equally well on the unseen data and outcomes are as expected, we finalize the model, else repeat steps 6, 7, and 8 until we find the most optimum model. Cross validation is a more popularly used technique in machine learning for model testing and validation

Prediction and Insights

There are two ways predictive modeling can be leveraged. One is the prediction that directly impacts the business. On the other hand, models also generate insights and hidden patterns that are useful for making data-driven decisions and business strategies.

Deployment or Productionization and Model Maintenance

Taking projects from Jupyter notebook to live is an important step to leveraging AI in real time and realizing its impact. Cloud giants like Microsoft Azure, Google Cloud Platform (GCP), and Amazon Web Services (AWS) provide deployment frameworks and MLOps components for easy productionization.

Also, you can have multiple problem statements which produce multiple machine learning solutions. Maintaining these models, versioning, updating with new data plays a vital role. To achieve this, a model maintenance framework and AI governance should be in place to obtain the highest impact.

Business Context

The business context plays a vital role in each step of a storyline. Figure 1-20 to Figure 1-30 demonstrate the role of a business context in each step of the life cycle.

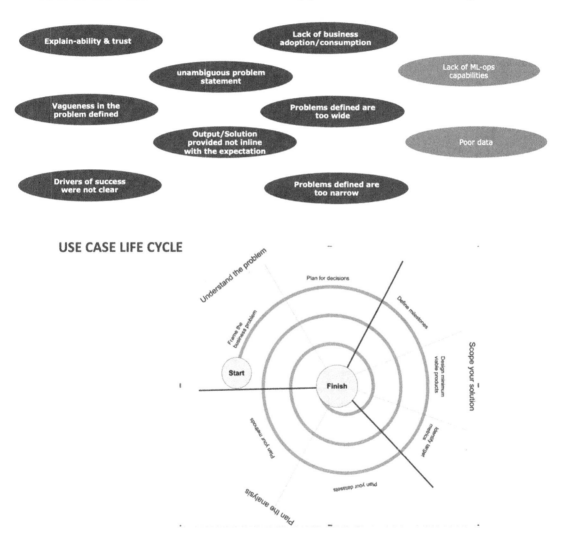

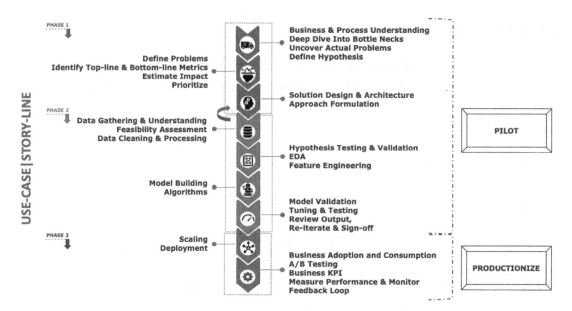

PHASE 1 | STEP 1

To understand the business process and define the **right** problem and avoid vagueness, we follow 6W-1H method

- ➤ **Get to the depth and breadth of the business, process and problems**
- ➤ **Understand the bottlenecks and scope for improvements**
- ➤ **Understand the GAP by understanding the current state and desired end state**
- ➤ **The GAP between them will help us define the problem and hypothesis**
- ➤ **Get concrete**
- ➤ **Focus on pain points**
- ➤ **Look for opposites**
- ➤ **Look for opposites**

How to do it? We must be able to ask ourselves a lot of questions, more importantly: **the right questions**.
The Golden Rule to define a project goal is to ask and refine "sharp" questions that are relevant, specific, and unambiguous; "How can I increase my profit?" is not a good question for any machine learning solution, "which kind of car in my fleet is going to fail first?" or "How much energy my production plant will consume in the next quarter?" are stronger examples of sharp questions.

PHASE 1 | STEP 2

Define bottom line and top line metrics & KPI's, estimate the Impact. If there are multiple problems encountered while in the previous step. Then prioritize use-case based on the

Estimated Impact (Primary and auxiliary)	The Business Need	The Hierarchical Approach

What is the business strategy?

Retention : Build churn model, CLTV etc.

New Customer Acquisition : Lead generation, Marketing spend optimization, promo optimization etc.

PHASE 1 | STEP 3

Translate the problem into ML/DS approach, Build a logical flow diagram and architecture

Value

Complexity of the approach

Analytics or Data science approach to solve any business problem should bring value to the project. It doesn't matter whether the approach is simple or complex.

And value can only be quantified with the help of business context. Someone who is expert in business as well as data science is the right person to formulate data science approaches. An engineer will always do complex things.

PHASE 2 | STEP 4

Before we look for what data we have, Let us list down what all data would be relevant for the use-case from business point of view and then check list against its availability & feasibility assessment

> External (Macro-economical, etc.)
> Competitors
> Social Media
> Customer Demographers
> Transaction Details
> Behavior Details

> Creating data dictionary and ERD, this will help us validate our understanding of data & its usage in the business context

We might spend **lot of time** on a data which doesn't even makes sense to use.
We might spend less time and vomit few **critical data sources** thinking its of no use.

PHASE 2 | STEP 5

What features make more sense and are actionable from business strategies POV

- **Hypothesis testing outcomes : alignment with business**
- **Remove the correlated variables (not statistically)**
- **Build more actionable features which can be implemented**

Few examples :
- Does sending coupon decrease churn
- Females has higher churn rate
- Early adopters churn is less

Research on customer behavior (what is driving sales > non-influenceable and influenceable drivers) and possible actions (e.g., discounts, visits, promotion material) and analyses if current data is capturing that behavior

Experimentations and iterations to see if we have right things to solve the given problem

PHASE 2 | STEP 6 & 7

Business context help us build the right model and fine tune

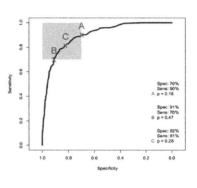

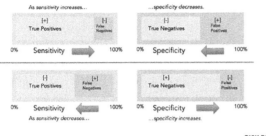

PHASE 3 | STEP 9

Adoption Strategy & Feedback Loop

Close stakeholder and realistic expectation management, help business to build solution (high transparency)	*Setup of stepwise solution building and testing process*	*Gain trust from business stakeholders*
Experienced guide through project	*Hypotheses need to be created with business and aligned with existing data points*	*Explain-ability*

 Significant aspect is the ability to pass on the results given by the data. People naturally have biased opinions that affect how they perceive results; we must find the most effective way to "tell the story" about the data; this is a highly relevant step in a project success.

 Explain-ability : Understanding the WHY part of model predictions is vital, and it should be in line the business process and hypothesis validated. Models are machines and they can go wrong. So, explain ability of these models are crucial before deploying them.

CHAPTER 2

Product360: Sentiment and Emotion Detector

This chapter explores sentiment analysis and emotion detection along with text preprocessing and feature engineering methods. Also, we are exploring different machine learning techniques to train classifier models and evaluate using a confusion matrix. In the second half of the chapter, let's pull real-time data from Twitter, and predict sentiment and emotions to generate insights. Also, we generate a script for automated reporting, which sends reports to a given set of e-mail addresses. This chapter gives you an overall picture of implementing an end-to-end pipeline that provides powerful insights about any product available on the market.

This chapter covers the following topics.

- Problem statements

- Building an emotion classifier model

- Label encoding

- Train-test split

- Feature engineering

- Model building

- Confusion matrix for a selected model

- Real-time data extraction

- Generating sentiment and emotions

- Visualizations and insights

- Automated report generation

© Akshay Kulkarni, Adarsha Shivananda and Anoosh Kulkarni 2022
A. Kulkarni et al., *Natural Language Processing Projects*, https://doi.org/10.1007/978-1-4842-7386-9_2

Problem Statement

Product360 is an end-to-end solution that helps users understand consumer sentiment on a product or service. For example, input a product name, such as Nokia 6.1, and decode 360-degree customer and market behavior. It is a simple solution that can solve multiple business problems and help with making business strategies.

Sentiment analysis involves finding the sentiments score for a given sentence. It categorizes a sentence as positive, negative, or neutral. It captures the public sentiment in reaction to a product or brand, which influences future business decision-making. However, this approach can only classify the texts into a positive, negative, or neutral class. This is not enough method when there are so many possible emotions attached to it. The desired solution to the problem is emotion detection along with sentiment.

Emotion detection involves identifying emotions (sad, angry, happy, etc.) from sentences. Data is extracted from social media like Twitter and Facebook, and e-commerce websites, and processed and analyzed using different NLP and machine learning techniques that provide the 360-degree view of that product, enabling better decision-making.

Approach Formulation

There are multiple sentiment prediction libraries, but we do not have the same for emotion detection, which is more robust. We build an emotion classifier as a part of this exercise. And using the classifier we built and sentiment prediction libraries, we predict the emotion and sentiment of a product using Twitter data. The detailed report is sent to the business team through an automated e-mail.

Data is collected from Twitter by using Twitter search API by specifying the search key. So, tweets related to any product can be used as the testing data. It is also possible to collect date and geolocation data.

Figure 2-1 is a flowchart that explains how the product works at a high level.

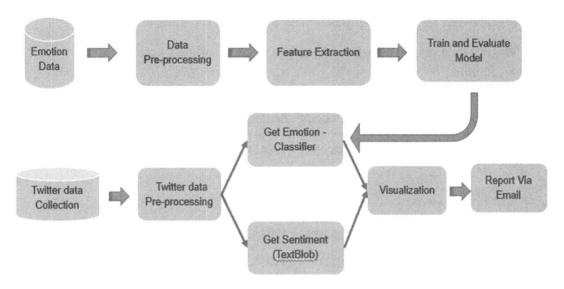

Figure 2-1. *Flowchart*

Steps to Solve This Problem

1. Build an emotion classifier model

2. Twitter data extraction

3. Data preprocessing

4. Emotion prediction

5. Sentiment prediction

6. Visualization and insights

7. Send report via mail

Building an Emotion Classifier Model

This model helps you understand the emotions around given sentences and reviews.

Data for Emotion Classifiers

To train the emotion classifier, let's use the ISEAR data set. Download the data from this book's git link.

- The data set has 7666 phrases classified into seven basic emotions.
- The data set classifies joy, anger, fear, disgust, sadness, guilt, and shame.

Emotion	Sentence
Anger	1094
Disgust	1094
Shame	1094
Sadness	1094
Fear	1093
Joy	1092
Guilt	1091
Total	**7652**

The data set consists of the following two columns.

- EMOTION lists the emotions.
- TEXT features the number of corresponding sentences.

Let's import the data set.

```
import pandas as pd
data = pd.read_csv('ISEAR.csv')
data.columns =['EMOTION', 'TEXT']
data.head()
```

Figure 2-2 shows the output of first five rows.

	EMOTION	TEXT
0	joy	[On days when I feel close to my partner and ...
1	fear	Every time I imagine that someone I love or I ...
2	anger	When I had been obviously unjustly treated and...
3	sadness	When I think about the short time that we live...
4	disgust	At a gathering I found myself involuntarily si...

Figure 2-2. *Output*

Data Cleaning and Preprocessing

Data cleaning is important to obtain better features and accuracy. You can achieve this by doing text preprocessing steps on the data.

The preprocessing steps are as follows.

1. Lowercase

2. Remove special characters

3. Remove punctuation

4. Remove stop words

5. Correct spelling

6. Normalization

The following are the libraries to preprocess the text. NLTK is a predominant free source Python package for text preprocessing.

```
#Importing the libraries for building Emotion Classifier
import pandas as pd
from nltk.corpus import stopwords
from nltk.stem.wordnet import WordNetLemmatizer
import string
from textblob.classifiers import NaiveBayesClassifier
from textblob import TextBlob
from nltk.corpus import stopwords
from nltk.stem import PorterStemmer
```

```
from textblob import Word
from nltk.util import ngrams
import re
from nltk.tokenize import word_tokenize
import matplotlib.pyplot as plt
from sklearn.feature_extraction.text import CountVectorizer,TfidfVectorizer
from sklearn.decomposition import LatentDirichletAllocation
import sklearn.feature_extraction.text as text
from sklearn.decomposition import NMF, LatentDirichletAllocation,
TruncatedSVD
from sklearn import model_selection, preprocessing, linear_model,
naive_bayes, metrics, svm
import xgboost
from sklearn import decomposition, ensemble
import pandas, numpy, textblob, string
import re
import nltk
from sklearn.metrics import classification_report
from sklearn.metrics import confusion_matrix
from sklearn.metrics import accuracy_score
from sklearn.metrics import mean_absolute_error
```

Let's look at more detail in the preprocessing steps and learn how to implement them.

1. Convert uppercase letters to lowercase.

    ```
    data['TEXT'] = data['TEXT'].apply(lambda a: "
    ".join(a.lower() for a in a.split()))
    ```

2. Remove white space and special characters.

    ```
    data['TEXT'] = data['TEXT'].apply(lambda a: " ".join(a.
    replace('[^\w\s]','') for a in a.split()))
    ```

3. Remove the stop words.

    ```
    stop = stopwords.words('english')
    ```

```
data['TEXT'] = data['TEXT'].apply(lambda a: " ".join(a for
a in a.split() if a not in stop))
```

4. Correct spelling.

```
data['TEXT'] = data['TEXT'].apply(lambda a:
str(TextBlob(a).correct()))
```

5. Do stemming.

```
st = PorterStemmer()
data['TEXT'] =  data['TEXT'].apply(lambda a: "
".join([st.stem(word) for word in a.split()]))
```

After completing all the preprocessing steps, this is what the data looks like.

```
data.head()
```

Figure 2-3 shows the output after preprocessing.

	EMOTION	TEXT
0	joy	days feel close partner friends á feel peace a...
1	fear	every time imagine someone love could contact ...
2	anger	obviously unjustly treated possibility á eluci...
3	sadness	think short time live relate á periods life th...
4	disgust	gathering found involuntarily sitting next two...

Figure 2-3. *Output*

Label Encoding

The target encoding is an approach to convert categorical value to numerical value. There are seven categories in this data, and we must encode them to proceed further. We are using the label encoder function to encode these categories.

The following shows the data before encoding.

```
data['EMOTION'].value_counts()
```

Figure 2-4 shows the output before encoding.

```
anger      1094
shame      1094
sadness    1094
disgust    1094
fear       1093
joy        1092
guilt      1091
```

Figure 2-4. *output*

Labels encode the target variable

```
object = preprocessing.LabelEncoder()
data['EMOTION'] = object.fit_transform(data['EMOTION'])
```

The following is the data after encoding.

```
data['EMOTION'].value_counts()
```

Figure 2-5 shows the output after encoding.

```
0    1094
1    1094
5    1094
6    1094
2    1093
4    1092
3    1091
```

Figure 2-5. *Output*

Train-Test Split

The data is split into two parts: one part trains the model, which is the training set, and the other part evaluates the model, which is the test set. The train_test_split library from sklearn.model_selection is imported to split the data frame into two parts.

```
#train-test split
Xtrain, Xtest, Ytrain, Ytest = model_selection.train_test_
split(data['TEXT'], data['EMOTION'],stratify= data['EMOTION'])
```

Now that you have completed the train-test split step, the next step is to extract the features out of these texts. For this, we use two important methods.

Feature Engineering

Feature engineering is the process of creating a new feature considering the domain context. Let's implement the count vectorizer and TF-IDF techniques to obtain the relevant features from the data sets.

For more information about the count vectorizer and TF-IDF, please refer to Chapter 1.

```
cv = CountVectorizer()
cv.fit(data['TEXT'])

cv_xtrain =  cv.transform(Xtrain)
cv_xtest =  cv.transform(Xtest)
```

The following is word-level TF-IDF.

```
tv = TfidfVectorizer()
tv.fit(data['TEXT'])
```

Transform the training and validation data using TF-IDF object.

```
tv_xtrain =  tv.transform(Xtrain)
tv_xtest =  tv.transform(Xtest)
```

Now let's get into one of the crucial steps to build the multiclass text classification model. We explore the different algorithms in this section.

Model Building Phase

In this phase, we build different models using both count vectors and word-level TF-IDF as features, and then the model is finalized based on the accuracy level of the classifier.

Let's build a classifier function so that you can play around with the different algorithms.

```
def build(model_initializer, independent_variables_training, target,
independent_variable_test):
```

```
# fit
model_initializer.fit(independent_variables_training, target)
# predict
modelPred=classifier_model.predict(independent_variable_test)
return metrics.accuracy_score(modelPred, Ytest)
```

Let's use the preceding function and try various algorithms.

Multinomial Naive Bayes

The multinomial naive Bayes algorithm essentially calculates the probability of each category using the Bayes theorem. For more information, refer to Chapter 1.

Let's build a naive Bayes model.

The following uses naive Bayes generated with count vectors.

```
output = build(naive_bayes.MultinomialNB(), cv_xtrain, Ytrain, cv_xtest)
print(output)
```

The following uses naive Bayes generated with word-level TF-IDF vectors.

```
output = build(naive_bayes.MultinomialNB(), tv_xtrain, Ytrain, tv_xtest)
print(output)
```

```
#Output:
0.561944
0.565081
```

56.1% accuracy is obtained from count vectorizer features.

56.5% accuracy is obtained from TD-IDF vectorizer features.

Linear Classifier/Logistic Regression

For more information on the algorithm, please refer to Chapter 1.

The following builds a logistic regression model.

```
# for CV
output = build(linear_model.LogisticRegression(), cv_xtrain, Ytrain,
cv_xtest)
print(output)
```

```
# for TF-IDF
output = build(linear_model.LogisticRegression(), tv_xtrain, Ytrain,
tv_xtest)
print(output)

#Output:
0.565081
0.590172
```

Support-Vector Machine

For more information, refer to Chapter 1.

Let's build the SVM model.

```
#for cv
output = build(svm.SVC(), cv_xtrain, Ytrain, cv_xtest)
print(output)

#for TF-IDF
output = build(svm.SVC(), tv_xtrain, Ytrain, tv_xtest)
print(output)

#Output:
0.545739
0.578672
```

Random Forest

You learned how random forest works in chapter 1. It's an ensemble technique that consists of decision trees. Let's look at how this performs on the data.

The following builds a random forest model.

```
#for CV
output = build(ensemble.RandomForestClassifier(), cv_xtrain, Ytrain,
cv_xtest)
print(output)
```

```
#for TF-IDF
output = build(ensemble.RandomForestClassifier(), tv_xtrain, Ytrain,
tv_xtest)
print(output)

#Output:
0.553580
0.536330
```

Model Evaluation and Comparison Summary

We tried a few different machine learning algorithms using both count vectorizers and TF-IDF vectorizers. Table 2-1 shows the results. We considered accuracy in this case. You can consider other metrics like AUC, specificity, and F1 score for better evaluation.

Table 2-1. *Output Summary*

Algorithm	Feature Engineering	Accuracy
Naive Bayes	Count Vector	56%
	TF-IDF (Word Level)	57%
Linear Classifier	Count Vector	57%
	TF -IDF (Word Level)	**59%**
SVM	Count Vector	54%
	TF-IDF (Word Level)	57%
Random Forest	Count Vector	55%
	TF-IDF (Word Level)	53%

Since the linear classifier with word-level TF-IDF provides greater accuracy, let's select that combination in further steps.

Please note the accuracy can be improved with more data and computation power. However, the objective of this chapter is not to gain more accuracy but to understand and implement end to end.

Confusion Matrix for the Selected Model

Now let's evaluate and validate the model using a confusion matrix. Here we use classification_report, confusion_matrix, and accuracy_score from the sklearn library.

```
classifier = linear_model.LogisticRegression().fit(tv_xtrain, Ytrain)
val_predictions = classifier.predict(tv_xtest)

# Precision , Recall , F1 - score , Support
y_true, y_pred = Ytest, val_predictions
print(classification_report(y_true, y_pred))
print()
```

Figure 2-6 shows the detailed classification report.

	precision	recall	f1-score	support
0	0.47	0.51	0.49	261
1	0.60	0.66	0.63	276
2	0.66	0.69	0.68	252
3	0.50	0.47	0.48	279
4	0.66	0.75	0.70	277
5	0.69	0.61	0.65	278
6	0.56	0.45	0.50	290
accuracy			0.59	1913
macro avg	0.59	0.59	0.59	1913
weighted avg	0.59	0.59	0.59	1913

Figure 2-6. *Output*

The F1 score is not bad for a few of the target categories.

Now let's start with real-time data extraction.

Real-time Data Extraction

You want to make a product as real time as possible. To understand the sentiment and emotion of any product, you need public data that is freely available. If you have a retail website, you can fetch data from there. But that's very rare. And you always sell products on someone else's platform. The following are possible ways to collect data.

- Twitter

- Other social media like Facebook and LinkedIn

- E-commerce websites like Amazon

- Company websites

- News articles on companies or products

Getting data from all these data sources is the biggest task. We start with Twitter for this chapter.

All kinds of people share their opinion on Twitter about newly released products. It's one of the right platforms to fetch the data and analyze it. It also provides APIs to collect the data for any given search. You need to create the Twitter developer account, and we are good to use the tweets. Let's explore how to use this API and pull data.

Twitter API

To use this script, you should register a data-mining application with Twitter. After doing this, you are provided with a unique consumer key, consumer secret, access token, and access secret. Using these keys, you can use the functions from the Tweepy library to pull the data from Twitter without much hassle.

To get the real-time Twitter data, use the twitter_data_extraction.py file, consisting of all the required functions with necessary comments.

Run twitter_data_extraction.py from the Git source to extract Twitter data (see `https://github.com/agalea91/twitter_search`).

By executing all the functions from the .py file, you get the output shown in Figure 2-7. The data collected from Twitter is saved in JSON format by creating a separate directory. Figure 2-7 shows the snapshot of Samsung data extracted from Twitter.

```
Search phrase = samsung
search limit (start/stop): 2018-11-30 23:59:59
search limit (start/stop): 2018-11-29 23:59:59
max id (starting point) = 1068655917263142912
since id (ending point) = 1068293529397547009
count = 1
found 91 tweets
found 9 tweets
count = 2
found 100 tweets
count = 3
found 100 tweets
count = 4
```

Figure 2-7. *Output*

We are converting JSON to CSV and importing the CSV as a data frame to use this data.

```
twt=pd.read_json('/path/samsung/samsung_date.json',lines=True,orient='records')
```

We are only selecting dates and comments which are required for future analysis.

```
twt = twt[[ 'created_at','text']]
```

The tweets related to the Samsung brand are collected from the Twitter search API and used as the test set. It consists of 1723 sentences. The data set consists of two columns.

- **created_at** consists of date and time corresponding to tweets.

- **text** consists of corresponding tweets.

```
twt.tail()
```

Figure 2-8 shows the data extracted from Twitter.

	created_at	text
20492	2021-06-13 00:01:10+00:00	I know you wanna Facetime, baby, I have Samsun...
20493	2021-06-13 00:01:07+00:00	RT @femi_tm: @pepe_19i @sxsway Pepe_19i trying...
20494	2021-06-13 00:00:58+00:00	@FiscalDeGuerra Parcerias serão feitas com a i...
20495	2021-06-13 00:00:53+00:00	Retweet for Thomas the Tank Engine\nLike for S...
20496	2021-06-13 00:00:42+00:00	Samsung \n\n#hyunjinblueprint\n#hyunjinbubble\...

Figure 2-8. *Output*

The next step is preprocessing the tweets again so that you can avoid noise present in the text.

```
#text preprocess for twitter data
twt['text'] = twt['text'].str.lstrip('0123456789')
#lower casing
twt['text'] = twt['text'].apply(lambda a: " ".join(a.lower() for a in
a.split()))
#remove punctuation
twt['text'] = twt['text'].str.replace('[^\w\s]','')
#remove stopwords
sw = stopwords.words('english')
twt['text'] = twt['text'].apply(lambda a: " ".join(a for a in a.split() if
a not in sw))
#spelling correction
twt['text'].apply(lambda a: str(TextBlob(a).correct()))
```

After preprocessing

```
twt.tail()
```

Figure 2-9 shows the output after preprocessing.

	created_at	text
20492	2021-06-13 00:01:10+00:00	know wanna facetime baby samsung met mom told ...
20493	2021-06-13 00:01:07+00:00	rt femi_tm pepe_19i sxsway pepe_19i trying hos...
20494	2021-06-13 00:00:58+00:00	fiscaldeguerra parcerias serão feitas com inic...
20495	2021-06-13 00:00:53+00:00	retweet thomas tank engine like samsung girl h...
20496	2021-06-13 00:00:42+00:00	samsung hyunjinblueprint hyunjinbubble hyunjin...

Figure 2-9. *Output*

Predicting the Emotion

We extracted data using the Twitter API and collected enough to make predictions that led to some insights. After preprocessing, we predicted emotions using the function or algorithm built in the previous section. Then, we finalized a linear model, which gave better results. Finally, we saved the final data in a data frame.

Now, let's extract the relevant feature using the TF-IDF vectorizer. After that, we use the predict function to obtain the emotion.

```
Xpredict = twt['text']

# word level tf-idf
predict_tfidf = tv.transform(Xpredict)

# Get Predicted Emotion
twt['Emotion'] = classifier.predict(predict_tfidf)

twt.tail()
```

Figure 2-10 shows the output of emotion predictions.

	created_at	text	Emotion
20492	2021-06-13 00:01:10+00:00	know wanna facetime baby samsung met mom told ...	4
20493	2021-06-13 00:01:07+00:00	rt femi_tm pepe_19i sxsway pepe_19i trying hos...	0
20494	2021-06-13 00:00:58+00:00	fiscaldeguerra parcerias serão feitas com inic...	1
20495	2021-06-13 00:00:53+00:00	retweet thomas tank engine like samsung girl h...	1
20496	2021-06-13 00:00:42+00:00	samsung hyunjinblueprint hyunjinbubble hyunjin...	1

Figure 2-10. *Output*

Predicting the Sentiment

We predicted emotion using the function that we built earlier. But to predict sentiment, we are using a pretrained model. Let's use the *sentiment* function from TextBlob. We predict sentiment for each tweet and then aggregate to generate the insights.

The *sentiment* function has two outputs: *polarity* and *subjectivity*.

We concentrate on polarity, which provides the sentiment of that tweet on a scale of –1 to +1. You need to decide the threshold to get the final sentiment. For this exercise,

- if the polarity is > 0, then the positive sentiment

- if the polarity is < 0, then the negative sentiment

- if the polarity is = 0, then neutral sentiment

The following code provides the sentiments against the given input.

```
twt['sentiment'] = twt['text'].apply(lambda a: TextBlob(a).sentiment[0] )

def function (value):
    if value['sentiment'] < 0 :
        return 'Negative'
    if value['sentiment'] > 0 :
        return 'Positive'
    return 'Neutral'

twt['Sentiment_label'] = twt.apply (lambda a: function(a),axis=1)

twt.tail()
```

Figure 2-11 shows the output of sentiment predictions.

	created_at	text	Emotion	sentiment	Sentiment_label
20492	2021-06-13 00:01:10+00:00	know wanna facetime baby samsung met mom told ...	4	0.500000	Positive
20493	2021-06-13 00:01:07+00:00	rt femi_tm pepe_19i sxsway pepe_19i trying hos...	0	0.000000	Neutral
20494	2021-06-13 00:00:58+00:00	fiscaldeguerra parcerias serão feitas com inic...	1	0.000000	Neutral
20495	2021-06-13 00:00:53+00:00	retweet thomas tank engine like samsung girl h...	1	0.000000	Neutral
20496	2021-06-13 00:00:42+00:00	samsung hyunjinblueprint hyunjinbubble hyunjin...	1	0.000000	Neutral

Figure 2-11. *Output*

Visualization and Insights

We generated emotion and sentiment in the last section on Samsung data pulled from Twitter using APIs. But building the algorithms, predicting emotion and sentiment won't solve the purpose. Instead, you need insights so that businesses can make decisions on that.

This section discusses how to do that using different libraries. The output of the predictive models is visualized through the different graphs. We are leveraging libraries like Plotly and Matplotlib.

First, let's visualize the sentiments. This is done by a pie graph, which gives detailed insight. The cufflinks library generates the graph.

```
import chart_studio.plotly as py
import plotly as ply
import cufflinks as cf
from plotly.graph_objs import *
from plotly.offline import *

init_notebook_mode(connected=True)
cf.set_config_file(offline=True, world_readable=True, theme='white')

Sentiment_df = pd.DataFrame(twt.Sentiment_label.value_counts().reset_
index())
Sentiment_df.columns = ['Sentiment', 'Count']

Sentiment_df = pd.DataFrame(Sentiment_df)

Sentiment_df['Percentage'] = 100 * Sentiment_df['Count']/ Sentiment_
df['Count'].sum()
```

```
Sentiment_Max = Sentiment_df.iloc[0,0]
Sentiment_percent = str(round(Sentiment_df.iloc[0,2],2))

fig1 = Sentiment_df.iplot(kind='pie',labels='Sentiment',values='Count',text
info='label+percent', title= 'Sentiment Analysis', world_readable=True,
                    asFigure=True)

ply.offline.plot(fig1,filename="Sentiment")
```

Figure 2-12 shows the output of sentiment represented in a pie chart.

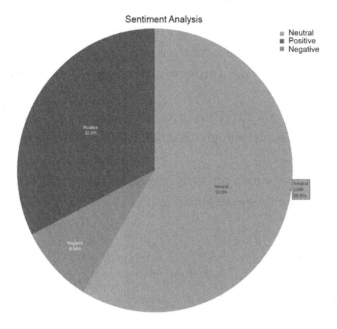

Figure 2-12. *Output*

The graph clearly shows that 58% of total comments are neutral and that 32% of total comments are positive. Only about 9% are negative. From a business point of view, it's a pretty good result for Samsung.

You can generate the following graphs to understand more about products.

- Sentiment by date reveals at what point of time a product or company was not doing well.

- Generate word cloud by a segment that tells you what positive and negative factors are for that product.

As with the sentiment, you see what the emotion results look like.

Emotion Analysis

As with the sentiment, you need to understand how the emotion to draw some significant conclusions. You predicted the emotion of every comment in earlier steps. Let's use that and generate charts to understand it better.

A pie chart visualizes the emotions of the document. This graph is also generated using the cufflinks library.

```
import chart_studio.plotly as py
import plotly as ply
import cufflinks as cf
from plotly.graph_objs import *
from plotly.offline import *

init_notebook_mode(connected=True)
cf.set_config_file(offline=True, world_readable=True, theme='white')
Emotion_df = pd.DataFrame(twt.Emotion.value_counts().reset_index())
Emotion_df.columns = ['Emotion', 'Count']
Emotion_df = pd.DataFrame (Emotion_df)
Emotion_df['Percentage'] = 100 * Emotion_df['Count']/ Emotion_df['Count'].sum()

Emotion_Max = Emotion_df.iloc[0,0]
Emotion_percent = str(round(Emotion_df.iloc[0,2],2))
fig = Emotion_df.iplot(kind='pie', labels = 'Emotion', values =
'Count',pull= .2, hole=.2 , colorscale = 'reds', textposition='outside',
colors=['red','green','purple','orange','blue','yellow','pink'],
textinfo='label+percent', title= 'Emotion Analysis', world_readable=True,
asFigure=True)
ply.offline.plot(fig,filename="Emotion")
```

Figure 2-13 shows the output of emotions represented in a pie chart.

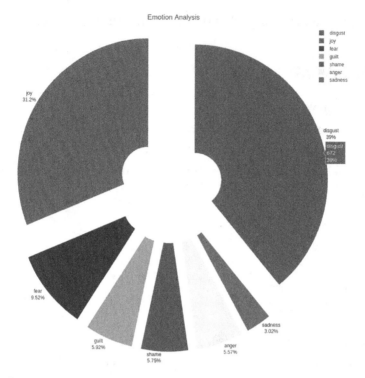

Figure 2-13. *Output*

The generated graph gives an emotional sense of the comments: 39% captured "disgust," 32% captured "joy," and the remaining captured fear, guilt, shame, anger, and sadness.

The good thing is 32% of the customers are happy. But you need to dig deeper to understand why 39% of comments are disgusting. There is chance customer were neutral with sentiments and disgusted.

To understand that, let's plot both sentiment and emotion on a single plot that answers all your questions.

Emotion and Sentiment Analysis

The stacked bar plot is the combination of both emotion and sentiment. It consists of each emotion and the corresponding sentiment on that emotion and gives the value count of the emotions.

The following generates a stacked bar plot chart.

```
import seaborn as sns
sns.set(rc={'figure.figsize':(11.7,8.27)})

Result = pd.crosstab(twt.Emotion, twt.Sentiment_label)
plt = Result.plot.bar(stacked=True,sort_columns = True)
plt.legend(title='Sentiment_label')
plt.figure.savefig('Emotion_Sentiment_stacked.png', dpi=400)
```

Figure 2-14 shows the output of sentiment and emotions represented in stacked bar chart.

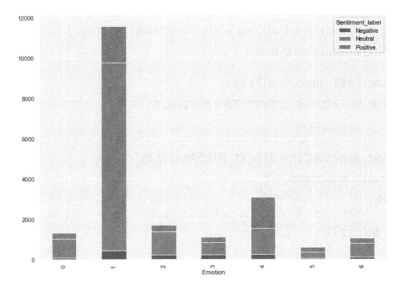

Figure 2-14. *Output*

This chart helps us analyze the overlap of semtiment and emotion in a better way.

Automated Reporting

We are nearing the final stage of the project, which is sending the reports to concerned business leaders. You need to create reports with charts and insights and must be sent to concerned people.

Another key aspect is it must be automated. You need to create an automated pipeline that pulls the data, processes it, and sends output to senior management.

As a part of that, an automated mailing system sends the mail automatically if you provide the e-mail addresses.

The output files like charts and tables are saved in the directory, and then they are sent via e-mails by executing the following script. The insights are sent as a report to the concerned persons via e-mail. For this, the user must specify the e-mail address and password along with the e-mail address of the concerned person(s). Import the smtplib library for sending automated e-mails.

```python
from email.mime.text import MIMEText
from email.mime.multipart import MIMEMultipart
import os

from email.mime.application import MIMEApplication
from email import encoders
import smtplib

def generate_email():

    dir_path = "Add PATH"
    files = ["Add files"]

    # Add concerned address (you can add multiple address also) and
    Password
    company_dict = ['xyz@gmail.com']
    password = "Password"

    for value in company_dict:
        # Add From email address
        From_address = 'From email id'
        To_address = value

        text = MIMEMultipart()
```

```
text['From'] = "xxxx"
text['To'] = To_address
text['Subject'] = "Emotion Detection and Sentiment Analysis Report"

body = " Hai \n Greetings of the day,\n We would like to inform
you that the data is more about, \n Emotion -  "+Emotion_Max+"
("+Emotion_percent+" %).\n Sentiment - " +Sentiment_Max+"
("+Sentiment_percent+" %).\n\n For the details please go through
the attachments bellow. \n\n\n\n\n Thank You."

text.attach(MIMEText(body, 'plain'))

for k in files:  # add files to the message

    file_location = os.path.join(dir_location, k)
    attachment = MIMEApplication(open(file_location, "rb").read(),
    _subtype="txt")
    attachment.add_header('Content-Disposition',obj, filename=k)
    text.attach(attachment)

smtp = smtplib.SMTP_SSL('smtp.gmail.com', 465)
smtp.login(From_address, password)
text1 = text.as_string()
smtp.sendmail(From_address, To_address, text1)
smtp.quit()

return 'Successfully Sent..!'

#Calling the function to send reports

generate_email ()
```

Note   To send mail via Gmail, you must allow "Less secure apps access" in your Google Account Security settings. This must be done explicitly due to the extra security issues from Gmail. For more information, go to `https://myaccount.google.com/security?pli=1#connectedapps`.

How is the solution different?

- The tool provides emotions with a sentiment.

- Insights of both emotion and sentiment along with the sentiment attached with emotions.

- An analysis report/insights are sent via mail along with the description, insights, and charts to those concerned.

Summary

- ISEAR data for training an emotion classifier.

- In the model-building phase, different behavior of classifiers is considered with different feature engineering mechanisms.

- Linear classifier with word-level TF-IDF feature engineering mechanism is taken into consideration by comparing the accuracy.

- Sentiment prediction using TextBlob, a pretrained model. Samsung data is collected from Twitter using Twitter API and predicts both emotion and sentiment.

- Insights are from three aspects: emotion analysis, sentiment analysis, and sentiment of emotion.

- The analysis report/insights are sent via e-mail to those concerned.

## CHAPTER 3

TED Talks Segmentation and Topics Extraction Using Machine Learning

TED Talks are knowledge videos of talks by experts in technology, entertainment, and design (hence TED). These conferences are arranged across the world. Great speakers come forward and share their experiences and knowledge. These talks are limited to a maximum length of 18 minutes and cover a wide range of topics. The videos are stored, and every video has a description of the video content.

This chapter groups TED Talks based on their description and uses various clustering techniques like k-means and hierarchical clustering. Topic modeling using Latent Dirichlet Allocation (LDA) is used to understand and interpret each cluster. The libraries used include Genism, NLTK, scikit-learn, and word2vec.

Problem Statement

There are tons of these videos across the world. They are recorded in different locations, with different topics and a variety of speakers. Tagging these videos manually to a certain category is the biggest challenge given the number of videos we have already. Let's look at leveraging machine learning and natural language processing to solve this problem.

The challenge here is that we don't have labeled data to train a classifier and predict the categories. Instead, we need to take the unsupervised learning approach. Let's look at how to formulate the approach to get the categories for each video.

© Akshay Kulkarni, Adarsha Shivananda and Anoosh Kulkarni 2022
A. Kulkarni et al., *Natural Language Processing Projects*, https://doi.org/10.1007/978-1-4842-7386-9_3

Approach Formulation

The approach has two parts to solve the preceding problem.

The first problem is grouping the documents. We need to use document clustering to segment the TED Talks. The document clustering technique has mainly two steps. First, convert the text to feature using different techniques like count vectorizer, TF-IDF, or word embeddings. Then using these features, perform clustering using k-means or hierarchical clustering.

Once the clustering is performed, the next problem is to understand the clusters. Then, you can perform topic modeling to extract topics and learn the cluster properties. Again, there are multiple algorithms to extract topics. We use LDA in this case.

Let's visualize the clusters and topic modeling results for a better understanding of TED Talks.

Figure 3-1 shows the logical flow diagram of the approach.

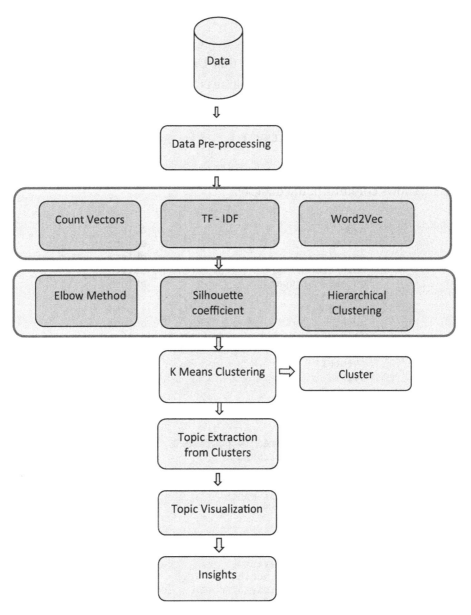

Figure 3-1. *Approach flow chart*

Data Collection

We are considering open source data for this. Download the data set from this book project's Gitlink.

Understanding the Data

We are considering the open source data for this. Download the data set from the repo where the code is located. The data set file name is 'Ted talks.csv'.

```python
import pandas as pd
ted_df = pd.read_csv('Ted talks.csv')
ted_df.dtypes
```

Figure 3-2 shows columns in the data set.

```
name                              object
title                             object
description                       object
main_speaker                      object
speaker_occupation                object
num_speaker                        int64
duration                           int64
event                             object
film_date                         object
published_date                    object
comments                           int64
tags                              object
languages                          int64
url                               object
views                              int64
transcript                        object
days_since_filmed_to_published     int64
days_since_published               int64
film_date.year                     int64
film_date.month                    int64
film_date.day-of-month             int64
film_date.day-of-week              int64
film_date.hour                     int64
film_date.minute                   int64
film_date.second                   int64
published_date.year                int64
published_date.month               int64
published_date.day-of-month        int64
published_date.day-of-week         int64
published_date.hour                int64
published_date.minute              int64
published_date.second              int64
dtype: object
```

Figure 3-2. *Output*

Now let's try to understand the data.

The output in Figure 3-2 defines the column name and the corresponding data types. All column names are self-explanatory.

```
ted_df.head()
```

Figure 3-3 shows the data head.

	name	title	description	main_speaker	speaker_occupation	num_speaker	duration	event	film_date	published_date	...	film_date.hour	film_
0	Ken Robinson: Do schools kill creativity?	Do schools kill creativity?	Sir Ken Robinson makes an entertaining and pro...	Ken Robinson	Author/educator	1	1164	TED2006	2006-02-25 01:00:00	2006-06-27 02:11:00	...	1	
1	Al Gore: Averting the climate crisis	Averting the climate crisis	With the same humor and humanity he exuded in ...	Al Gore	Climate advocate	1	977	TED2006	2006-02-25 01:00:00	2006-06-27 02:11:00	...	1	
2	David Pogue: Simplicity sells	Simplicity sells	New York Times columnist David Pogue takes aim...	David Pogue	Technology columnist	1	1286	TED2006	2006-02-24 01:00:00	2006-06-27 02:11:00	...	1	
3	Majora Carter: Greening the ghetto	Greening the ghetto	In an emotionally charged talk, MacArthur-winn...	Majora Carter	Activist for environmental justice	1	1116	TED2006	2006-02-26 01:00:00	2006-06-27 02:11:00	...	1	
4	Hans Rosling: The best stats you've ever seen	The best stats you've ever seen	You've never seen data presented like this. Wi...	Hans Rosling	Global health expert; data visionary	1	1190	TED2006	2006-02-22 01:00:00	2006-06-27 22:38:00	...	1	

Figure 3-3. *Output*

Check the number of rows and columns available in the data set.

```
ted_df.shape
```

The following is the output.

```
(2550,32)
```

The TED Talks file consists of 2550 rows and 32 columns. Out of that, only the transcript column is taken into consideration.

```
ted_df = ted_df[['title','transcript']].astype(str)
ted_df.head()
```

Figure 3-4 shows the data head.

	title	transcript
0	Do schools kill creativity?	Good morning. How are you?(Laughter)It's been ...
1	Averting the climate crisis	Thank you so much, Chris. And it's truly a gre...
2	Simplicity sells	(Music: "The Sound of Silence," Simon & Garfun...
3	Greening the ghetto	If you're here today — and I'm very happy that...

Figure 3-4. *Output*

Data Cleaning and Preprocessing

Cleaning text data is important to obtain better features. It can be achieved by doing some of the basic preprocessing steps on the data.

The following are the preprocessing steps.

1. Make words lowercase.

2. Remove stop words.

3. Correct spelling.

4. Remove numbers.

5. Remove whitespace and special characters.

First import all the required libraries.

```
import numpy as np
import nltk
import itertools
from nltk.tokenize import sent_tokenize, word_tokenize
import scipy
from scipy import spatial
import re
from textblob import TextBlob
sw = nltk.corpus.stopwords.words('english')
import gensim
```

```
from gensim.models import Word2Vec
from nltk.corpus import stopwords
from nltk.stem import PorterStemmer
from nltk.tokenize import word_tokenize
from nltk.stem.wordnet import WordNetLemmatizer
import string
from nltk.util import ngrams
from textblob import TextBlob
from textblob import Word
from nltk.stem.snowball import SnowballStemmer
stemmer = SnowballStemmer("english")
from nltk.stem import PorterStemmer
st = PorterStemmer()
from sklearn.feature_extraction.text import CountVectorizer,TfidfVectorizer
import gensim
from gensim.models import Word2Vec
from sklearn.cluster import KMeans
```

Let's create data cleaning and preprocessing function to perform necessary cleaning as discussed in Chapter 1.

```
def  text_processing(df):
    """"""=== Lower case ==="""
    df['transcript'] = df['transcript'].apply(lambda x: " ".join(x.lower()
    for x in x.split()))

    '''=== Removal of stop words ==='''
    df['transcript'] = df['transcript'].apply(lambda x: " ".join(x for x in
    x.split()if x not in sw))

    '''=== Spelling Correction === '''
    df['transcript'].apply(lambda x: str(TextBlob(x).correct()))

      '''=== Removal of Punctuation ==='''
    df['transcript'] = df['transcript'].str.replace('[^\w\s]', '')

    '''=== Removal of Numeric ==='''
    df['transcript'] = df['transcript'].str.replace('[0-9]', '')
```

```
    return df
```

```
ted_df = text_processing(ted_df)
print(ted_df)
```

Figure 3-5 shows the first few rows of the transcript column after completing the preprocessing steps.

```
                                                    transcript
0       good morn how are youlaughterit been great has...
1       thank much chri truli great honor have opportu...
2       music sound silenc simon garfunkelhello voic m...
3       if your here today im veri happi are youv all ...
4       about year ago took on task teach global devel...
5       thank have tell im both challeng excit my exci...
6       on septemb morn my seventh birthday came downs...
7       im go present three project rapid fire dont ha...
8       wonder be back love wonder gather must be wond...
9       im often ask what surpri about book say got wr...
10      im go take on journey veri quickli explain wis...
11      cant help but wish think about when your littl...
12      im luckiest guy world got see last case killer...
13      im realli excit be here today ill show some st...
14      ive been at mit for year went ted there onli o...
```

Figure 3-5. *Output*

For more information on stemming and lemmatization, please refer to Chapter 1.

```
ted_df['transcript'] = ted_df['transcript'].apply(lambda a: "
".join([s.stem(x) for x in a.split()]))
```

These preprocessing steps have been used in this work. One can include other steps also based on the requirement or based on the data.

Feature Engineering

In the NLP life cycle, feature engineering uses domain knowledge of the data to create features that make machine learning algorithms work. It is fundamental to the application of machine learning. We implement the following techniques to obtain the relevant features from data sets.

Count Vectors

For more information on count vectorizer and TF-IDF, please refer to Chapter 1.

```
cv =CountVectorizer()
cv.fit(ted_df['transcript'])
cv_tedfeatures = cv.transform(ted_df['transcript'])
```

TF-IDF Vectors

```
#word level TF-IDF
tv = TfidfVectorizer()
tv.fit(ted_df['transcript'])
tv_tedfeatures =  tv.transform(ted_df['transcript'])
```

Word Embeddings

Let's use the word2vec pretrained model from Gensim. Import and implement the same. Also, download the GoogleNews-vectors-negative300.bin pretrained model from www.kaggle.com/sandreds/googlenewsvectorsnegative300.

```
#Load
m1=gensim.models.KeyedVectors.load_word2vec_format('GoogleNews-vectors-
negative300.bin', binary=True)

# Function to get the embeddings
def get_embedding (x, out=False):
    if x in m1.wv.vocab:
        if out == True:
            return m1.wv.vocab[x]
        else:
            return m1[x]
    else:
        return np.zeros(300)

# Getting means
op =  {}
for i in ted_df['transcript']:
```

```
avg_vct_doc = (np.mean(np.array([get_embedding(a)for a in nltk.word_
tokenize((i))]), axis=0))
d = { i : (avg_vct_doc) }
op.update(d)
```

op

Figure 3-6 shows the mean of the embeddings.

```
1543,   0.02836929,   0.09126323,  -0.04454758,
          0.00049415,   0.03995693,  -0.06984885,   0.04054823,   0.05384475,
         -0.02971982,  -0.10276053,  -0.03801156,   0.01406198,  -0.08581767,
          0.07025365,   0.04345104,   0.08948038,   0.0011024 ,  -0.04239726,
         -0.02881136,   0.01687422,   0.04324164,  -0.01138773,   0.03274092,
         -0.01435886,  -0.07170025,   0.02306397,   0.03228389,  -0.02717961,
         -0.02511194,   0.02229446,  -0.05735477,  -0.0240358 ,  -0.00029277,
         -0.01531751,   0.03972094,  -0.00840874,   0.02287067,   0.05009453,
          0.05992658,  -0.03821913,   0.09209948,  -0.0147468 ,  -0.00837915,
         -0.01317505,  -0.0221782 ,  -0.00275255,   0.00242857,   0.03100006,
         -0.01238   ,   0.0475575 ,  -0.00815522,  -0.00893451,   0.00727858,
          0.01326992,  -0.04393348,  -0.03917407,   0.03777715,  -0.05380889,
```

Figure 3-6. *Output*

Let's separate the dict values and keys.

```
results_key = list()
results_value = list()
for key, value in op.items():
    results_key.append(key)
    results_value.append(np.array(value))
```

Figure 3-7 shows the output of word embeddings.

```
[array([ 0.02148347,  0.00941543,  0.02836929,  0.09126323, -0.04454758,
         0.00049415,  0.03995693, -0.06984885,  0.04054823,  0.05384475,
        -0.02971982, -0.10276053, -0.03801156,  0.01406198, -0.08581767,
         0.07025365,  0.04345104,  0.08948038,  0.0011024 , -0.04239726,
        -0.02881136,  0.01687422,  0.04324164, -0.01138773,  0.03274092,
        -0.01435886, -0.07170025,  0.02306397,  0.03228389, -0.02717961,
        -0.02511194,  0.02229446, -0.05735477, -0.0240358 , -0.00029277,
        -0.01531751,  0.03972094, -0.00840874,  0.02287067,  0.05009453,
         0.05992658, -0.03821913,  0.09209948, -0.0147468 , -0.00837915,
        -0.01317505, -0.0221782 , -0.00275255,  0.00242857,  0.03100006,
        -0.01238   ,  0.0475575 , -0.00815522, -0.00893451,  0.00727858,
         0.01326992, -0.04393348, -0.03917407,  0.03777715, -0.05380889,
        -0.00770878,  0.04582382, -0.04901921, -0.04435144, -0.01674103,
        -0.0340198 , -0.03636459,  0.07095297, -0.02960991,  0.06109828,
         0.031721  ,  0.04469927,  0.05684717,  0.02233426, -0.10884775,
        -0.03920979,  0.05851065,  0.07266818,  0.02193966,  0.07829539,
         0.00259239, -0.03937387,  0.04251901, -0.01220335, -0.04377269,
        -0.02670283, -0.06503243,  0.08960646,  0.00492547,  0.01778738,
         0.02628484   0.02073154   0.05896200   0.07278573   0.0221276
```

Figure 3-7. *Output*

We generated a single vector for each document in the corpus.

Now that we have implemented three types of feature engineering techniques. Let's build the models using all three features and observe which one performs better.

So, let's start the model building phase.

Model Building Phase

In this phase, we build clusters using the k-means method. To define an optimal number of clusters, we considered different methods like elbow, silhouette coefficient, and dendrogram method. All these methods are considered by using count vectors, word-level TF-IDF, and word embeddings as features, and then the model has been finalized based on the performance.

K-means Clustering

For more information on the k-means algorithm, please refer to Chapter 1.

There are various techniques to define an optimal number of clusters, including the elbow method, the silhouette score, and the dendrogram approach.

Elbow Method

The elbow method is a type of method for checking the consistency of clusters created. It finds the ideal number of clusters in data. Explained variance considers the percentage of variance explained and derives an ideal number of clusters. If the deviation percentage explained is compared with the number of clusters, the first cluster adds a lot of information, but at some point, explained variance decreases, which gives an angle on the graph. At the moment, the number of clusters is selected.

Silhouette Coefficient

The silhouette coefficient, or silhouette score, tells how much the object is similar to other clusters compared to its own cluster. The value varies from –1 to 1, where a high value indicates that the cluster fits well with itself, and the cluster does not match the neighboring cluster.

The silhouette value is calculated using distance metrics like Euclidean distance, the Manhattan distance, and so on.

Count Vectors as Features

Since we built the count vectorizer feature earlier, let's use that.

Elbow Method

Determine k using the elbow method.

```
from sklearn.cluster import KMeans
import matplotlib.pyplot as plt

elbow_method = {}
for k in range(1, 10):
    kmeans_elbow = KMeans(n_clusters=k).fit(cv_tedfeatures)
    elbow_method[k] = kmeans_elbow.inertia_
plt.figure()
plt.plot(list(elbow_method.keys()), list(elbow_method.values()))
plt.xlabel("Number of cluster")
plt.ylabel("SSE")
plt.show()
```

Figure 3-8 shows the elbow output.

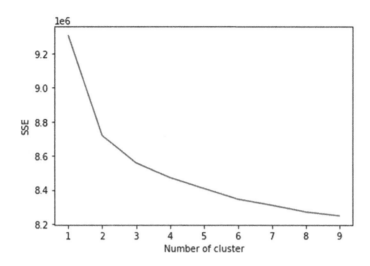

Figure 3-8. *Output*

Silhouette

```
from sklearn.metrics import silhouette_score
from sklearn.cluster import KMeans

for n_cluster in range(2, 15):
    kmeans = KMeans(n_clusters=n_cluster).fit(cv_tedfeatures)
    label = kmeans.labels_
    sil_coeff = silhouette_score(cv_tedfeatures, label, metric='euclidean')
    print("For n_clusters={}, The Silhouette Coefficient is
    {}".format(n_cluster, sil_coeff))
```

Figure 3-9 shows silhouette output.

```
For n_clusters=2, The Silhouette Coefficient is 0.2207550057687619
For n_clusters=3, The Silhouette Coefficient is 0.12756514226864543
For n_clusters=4, The Silhouette Coefficient is 0.11739901116141993
For n_clusters=5, The Silhouette Coefficient is 0.09339371908643401
For n_clusters=6, The Silhouette Coefficient is 0.07840920297133942
For n_clusters=7, The Silhouette Coefficient is 0.07551906276188601
For n_clusters=8, The Silhouette Coefficient is 0.07429325800675797
For n_clusters=9, The Silhouette Coefficient is 0.0446014420164Q7505
For n_clusters=10, The Silhouette Coefficient is 0.06158856323359691
For n_clusters=11, The Silhouette Coefficient is 0.04003568222803476
For n_clusters=12, The Silhouette Coefficient is 0.024753464380254785
For n_clusters=13, The Silhouette Coefficient is 0.022479722948913688
For n_clusters=14, The Silhouette Coefficient is 0.028816617432716466
```

Figure 3-9. *Output*

All the techniques suggest that two clusters are the ideal number of clusters.

Let's also implement other methods and see the results. We can leverage the same code used for the count vectorizer, change the input feature accordingly. Let's start with the output.

TF-IDF as Features

Elbow Method

Figure 3-10 shows a silhouette output.

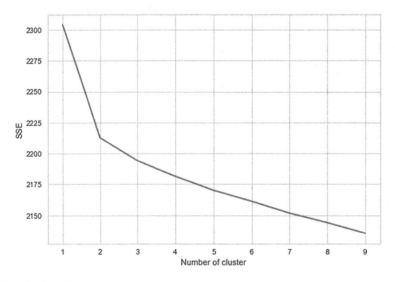

Figure 3-10. *Output*

Silhouette

Figure 3-11 shows a silhouette output.

```
For n_clusters=2, The Silhouette Coefficient is 0.21316025004986813
For n_clusters=3, The Silhouette Coefficient is 0.05022589300731797
For n_clusters=4, The Silhouette Coefficient is 0.03966218355177137
For n_clusters=5, The Silhouette Coefficient is 0.033681494041442925
For n_clusters=6, The Silhouette Coefficient is 0.029655396979534836
For n_clusters=7, The Silhouette Coefficient is 0.02491576793905632
For n_clusters=8, The Silhouette Coefficient is 0.0331495628491884
For n_clusters=9, The Silhouette Coefficient is 0.025834949451346043
For n_clusters=10, The Silhouette Coefficient is 0.029468074234528248
For n_clusters=11, The Silhouette Coefficient is 0.035592438939874634
For n_clusters=12, The Silhouette Coefficient is 0.026291452891953743
For n_clusters=13, The Silhouette Coefficient is 0.03079030155838663
For n_clusters=14, The Silhouette Coefficient is 0.023821759951954152
```

Figure 3-11. *Output*

Even though this method suggests two clusters, the data distribution between two clusters is not great. One cluster has very little data. Let's also build the clusters using word embeddings before making any further decisions.

Word Embeddings as Features

Here are the outcomes using word embeddings as features, and again code would remain the same as the one used for count vectorizer.

Elbow Method

Figure 3-12 shows a silhouette output.

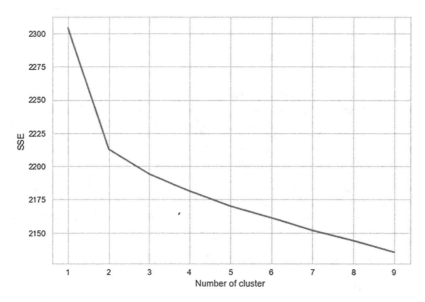

Figure 3-12. *Output*

Silhouette

Figure 3-13 shows a silhouette output.

```
For n_clusters=2, The Silhouette Coefficient is 0.098758633638018
For n_clusters=3, The Silhouette Coefficient is 0.056655051741800036
For n_clusters=4, The Silhouette Coefficient is 0.05749721023004869
For n_clusters=5, The Silhouette Coefficient is 0.05523648322541062
For n_clusters=6, The Silhouette Coefficient is 0.05073497536539663
For n_clusters=7, The Silhouette Coefficient is 0.054330075138415666
For n_clusters=8, The Silhouette Coefficient is 0.059333543270665645
```

Figure 3-13. *Output*

The optimal number of clusters is two. Let's use word2vec as the final text-to-features tool since it best captures semantics and context.

Building Clustering Model

Let's build a k-means clustering model. For more information on this, please refer to Chapter 1.

Directly jump to the implementation.

```
segments = KMeans(n_clusters=2)
segments.fit(results_value )
#segment outputs
output = segments.labels_.tolist()

ted_segmentaion =  {'transcript': results_key, 'cluster': output}
output_df = pd.DataFrame(ted_segmentaion)
#talks per segment
output_df['cluster'] = segments.labels_.tolist()

output_df['cluster'].value_counts()


Output:
0    1265
1    1200

cluster_1 = output_df[output_df.cluster == 1]
cluster_0 = output_df[output_df.cluster == 0]
```

Cluster Visualization

Each cluster is visualized through the monogram word cloud.

```
# cluster 1 visualization

from wordcloud import WordCloud, STOPWORDS

# Mono Gram

wordcloud = WordCloud(width = 1000, height = 500,collocations = False).
generate_from_text(' '.join(cluster_1['transcript']))
# Generate plot
plt.figure(figsize=(15,8))
plt.imshow(wordcloud)
plt.axis("off")
plt.show()
```

Figure 3-14 shows a cluster word cloud.

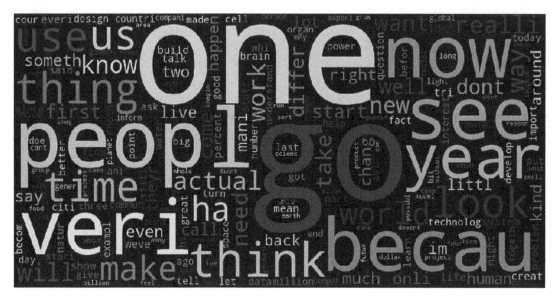

Figure 3-14. *Output*

```
# Similarly for segment 0 visualization
wordcloud = WordCloud(width = 1000, height = 500,collocations = False).
generate_from_text(' '.join(cluster_0['transcript']))

# Generate plot
plt.figure(figsize=(15,8))
plt.imshow(wordcloud)
plt.axis("off")
plt.show()
```

Figure 3-15 shows another cluster word cloud.

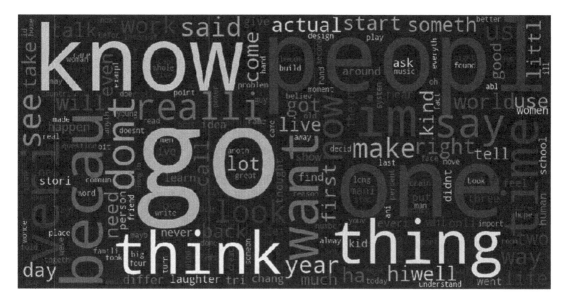

Figure 3-15. *Output*

We have successfully built the clusters and visualized them as well. Though we cannot interpret or conclude much from the word clouds, in ideal scenarios, word clouds are the best way to visualize text.

Next, let's perform topic modeling and further interpret the clusters.

Topic Modeling

Topic modeling is a method to derive topics from the corpus of text. A document may cover any number of topics, like cricket, entertainment, and so on. Using LDA, you can extract all such topics present in the document.

Let's start building one.

First let's create text processing function required for topic modeling.

```
def process(doc):
    toks = [w for s in nltk.sent_tokenize(doc) for w in nltk.word_
tokenize(s)]
    filt_toks = []
    for i in toks:
        if re.search('[a-zA-Z]', i):
            filt_toks.append(i)
```

```
    post_process = [st.stem(t) for t in filt_toks]
    return post_process
```

Topic Modeling for Cluster 1

```
#import
from enism import corpora, models, similarities

toks = [process(a) for a in cluster_1.transcript]

talks = [[x for x in y if x not in sw] for y in toks]

#dictionary from text
dictionary = corpora.Dictionary(talks)

#bow
doc = [dictionary.doc2bow(text) for text in talks]

#topic modeling
tm = models.LdaModel(doc, num_topics=5,
                            id2word=dictionary)

tm.show_topics()
```

Figure 3-16 shows the topic modeling output.

```
[(0,
  '0.007*"need" + 0.006*"countri" + 0.005*"percent" + 0.004*"dont" + 0.004*"citi" + 0.004*"water" + 0.004*"live" + 0.004*"start" + 0.004*"chang" + 0.003*"foo
d"'),
 (1,
  '0.005*"actual" + 0.005*"build" + 0.005*"space" + 0.004*"new" + 0.004*"earth" + 0.004*"light" + 0.004*"planet" + 0.004*"littl" + 0.004*"someth" + 0.003*"li
fe"'),
 (2,
  '0.005*"actual" + 0.004*"new" + 0.004*"data" + 0.004*"need" + 0.004*"dont" + 0.004*"say" + 0.004*"right" + 0.004*"design" + 0.004*"chang" + 0.003*"even"'),
 (3,
  '0.005*"war" + 0.005*"hi" + 0.004*"govern" + 0.004*"mani" + 0.003*"right" + 0.003*"polit" + 0.003*"state" + 0.003*"women" + 0.003*"even" + 0.003*"countr
i"'),
 (4,
  '0.007*"brain" + 0.006*"cell" + 0.006*"actual" + 0.005*"human" + 0.005*"differ" + 0.004*"bodi" + 0.004*"littl" + 0.003*"cancer" + 0.003*"theyr" + 0.003*"i
m"')]
```

Figure 3-16. *Output*

```
import pyLDAvis.gensim
pyLDAvis.enable_notebook()
import warnings
warnings.filterwarnings("ignore", category=DeprecationWarning)

pyLDAvis.gensim.prepare(tm, doc, dictionary)
```

Figure 3-17 shows a topic modeling visual.

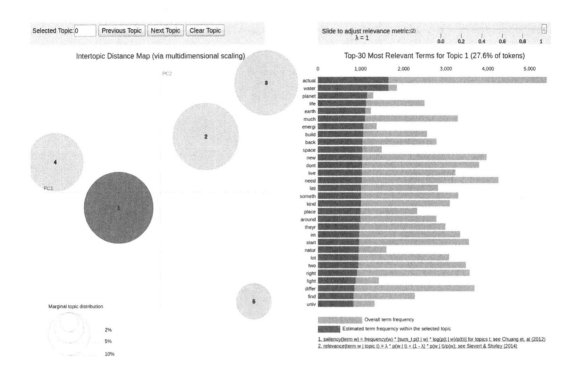

Figure 3-17. *Output*

Topic Modeling for Cluster 0

```
# dictionary
dictionary1 = corpora.Dictionary(talks)

# bow
doc1 = [dictionary1.doc2bow(text) for text in talks]

tm2 = models.LdaModel(doc1, num_topics=6,
                          id2word=dictionary)
tm2.show_topics()
```

Figure 3-18 shows the topic modeling output.

Figure 3-18. *Output*

```
pyLDAvis.gensim.prepare(tm2, doc1, dictionary1)
```

Figure 3-19 represents topic modeling visual for another cluster.

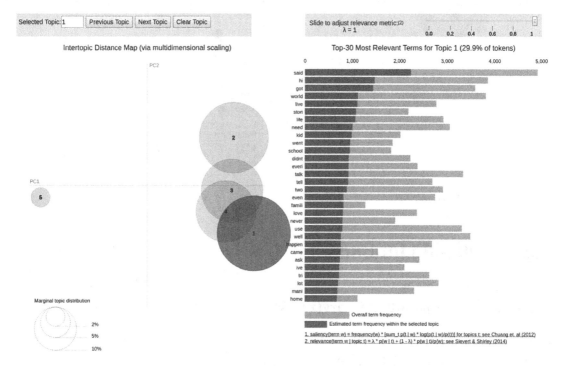

Figure 3-19. *Output*

We finally extracted the topics from both clusters and visualized them. This acts
as cluster properties as we analyze structured data that we can use for interpretation,
cluster naming, and business strategies.

It seems like the data set is not suitable for a document clustering task. Ideally, we would have more than two clusters, and clustering would segment so that each cluster topics and word cloud have relevant, unique topics in them.

Suppose you build a document clustering model to segment book documents for an online library into the respective categories. Let's say there are 10,000 such documents. These books can be of any type—technologies like Java/Python, legal books, comics, sports magazines, business books, and history books.

Similarly, as you learned in this chapter, we perform text processing and text-to-features and use these features for building clustering models using k-means or hierarchal algorithms after finding the optimal number of clusters. For example, there are seven clusters as output.

Finally, when we visualize them with word cloud and perform topic modeling. The following are the probable outcomes.

- Cluster 1 topics and word clouds have words like Python, *data science*, code, implement, project, OOPS, and programming

- Cluster 2 topics and word clouds have words like Virat Kholi, *cricket*, *football*, *top* runs, *goals*, and *wicket*

- Cluster 3 topics and word clouds have words like retail, marketing, management, profit, loss, and revenue

By looking at the topics and word clouds, you can conclude that cluster 1 mostly has technology-related topics, so tag all the cluster 1 documents accordingly. Cluster 2 has topics related to sports, so tag all the cluster 2 documents accordingly. Cluster 3 has documents related to business, so tag all the cluster 3 documents accordingly.

Conclusion

To summarize, the following were covered in this chapter.

- Understanding business problems

- Data collection from open source

- Understanding data and creating a separate data frame for another use case by considering only the column that consists of raw TED Talks

- Text preprocessing

- Feature engineering techniques like count vectorizer, word-level TF-IDF, word embeddings

- At the model building phase, k-means unsupervised learning approach

- Considered three methods: elbow, silhouette score, and dendrogram to define the number of clusters

- Clusters are visualized through monogram word cloud to get more sense out of it and for cluster interpretation

- Topic extraction from each cluster using LDA techniques to further enhance the cluster interpretation

- Each topic model for clusters is visualized separately by using pyLDAvis

- Insights are drawn from word clouds on clusters and their corresponding topic models

CHAPTER 4

Enhancing E-commerce Using an Advanced Search Engine and Recommendation System

A few decades ago, no one would have ever imagined that we could buy a 55-inch TV at midnight while sitting at home watching a 22-inch TV. Thanks to the Internet and e-commerce, we can buy any item at any time from anywhere, and it is delivered quickly. Flexibility has made e-commerce businesses expand exponentially. You don't have to visit the store, products have unlimited options, prices are lower, no standing in line to pay, and so forth.

Given the traction e-commerce is getting, many big names are taking part in it. Companies keep technology in check along with operations, supply chains, and marketing. To survive competition, the right use of digital marketing and social media presence is also required. Also, most importantly, businesses must leverage data and technology to personalize the customer experience.

Problem Statement

One of the most talked-about problems of this era is recommendation systems. Personalization is the next big data science problem. It's almost everywhere—movies, music, e-commerce sites, and more.

© Akshay Kulkarni, Adarsha Shivananda and Anoosh Kulkarni 2022
A. Kulkarni et al., *Natural Language Processing Projects*, https://doi.org/10.1007/978-1-4842-7386-9_4

Since the applications are wide, let's pick an e-commerce product recommendation as one of the problem statements. There are multiple types of recommender systems. But the one that deals with text is content-based recommender systems. For example, if you see the diagram shown in Figure 4-1, similar products have been recommended based on the product description of the clicked product. Let's explore building this kind of recommender system in this chapter.

On the same lines, search engines in e-commerce websites also play a major role in user experience and increasing revenue. The search bar should give the relevant matches for a search query, and wrong results eventually result in the churn of the customers. This is another problem that we plan to solve.

To summarize, in this project, we aim to build a search and recommender system that can search and recommend products based on an e-commerce data set.

Approach

Our main aim is to recommend the products or items based on users' historical interests. A recommendation engine uses different algorithms and recommends the most relevant items to the users. It initially captures the past behavior of the users. It recommends products based on that.

Let's discuss the various types of recommendation engines in brief before we move further. Figure 4-1 shows the types of recommendation engines.

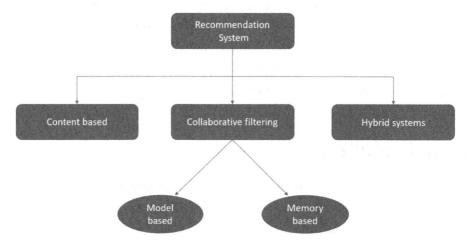

Figure 4-1. *Types of recommendation engines*

The following are various types of recommendation engines.

- Market basket analysis (association rule mining)

- Content based

- Collaborative filtering

- Hybrid systems

- ML clustering based

- ML classification based

- Deep learning and NLP based

Content-Based Filtering

Content Filtering algorithm suggests or predicts the items similar to the ones that a customer has liked or shown any form of interest. Figure 4-2 shows the example of content-based filtering.

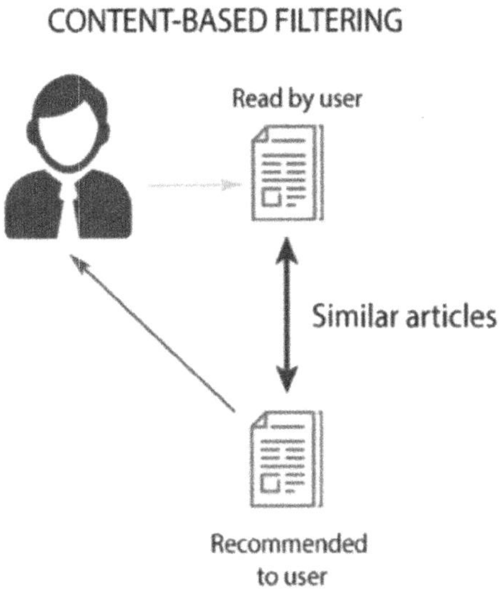

Figure 4-2. *Content-based filtering*

The project aims to use deep learning techniques for information retrieval rather than the traditional word comparison approach to get better results. Also, it focuses on recommender systems that are everywhere and creates personalized recommendations to increase the user experience.

The methodology involves the following steps.

1. Data understanding

2. Preprocessing

3. Feature selection

4. Model building

5. Returning search queries

6. Recommending product

Figure 4-3 shows the Term Frequency–Inverse Document Frequency (TF-IDF) vectors-based approach to building a content-based recommendation engine that gives a matrix where every word is a column.

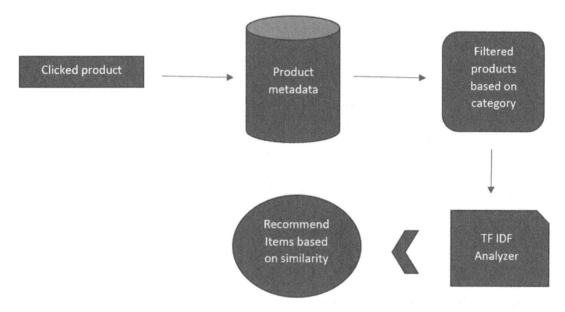

Figure 4-3. *TF-IDF-based architecture*

Figure 4-4 shows the architecture for a product search bar. Here, word embeddings are used. Word embedding is a language modeling technique where it converts text into real numbers. These embeddings can be built using various methods, mainly neural networks.

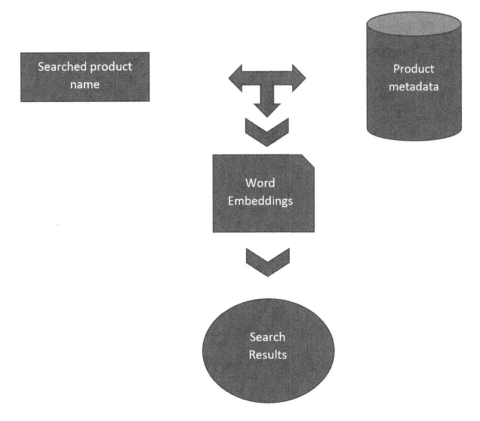

Figure 4-4. *Word embeddings-based architecture*

Environment Setup

Table 4-1 describes the environment setup that was used in this project. But, you could you Linux or macOS. To install Anaconda on Linux or macOS, please visit www.anaconda.com.

Table 4-1. *Environment Setup*

Set Up	Version
Processor	Intel(R) Core(TM) i5-4210U CPU @1.70GHz 2.40 GHz
Operating System	Windows 10- 64bit
Installed Memory (RAM)	8.00 GB
Anaconda Distribution	5.2.0
Python	3.6.5
Notebook	5.5.0
NumPY	1.14.3
pandas	0.23.0
scikit-learn	0.19.1
Matplotlib	2.2.2
Seaborn	0.8.1
NLTK	3.3.0
Gensim	3.4.0

Understanding the Data

The e-commerce product recommendation data set has 20,000 observations and 15 attributes. The 15 features are listed in Table 4-2.

Table 4-2. *Variables Present in the Data Set*

Attribute Name	Data Type
uniq_id	object
crawl_timestamp	object
product_url	object
product_name	object
Pid	object
retail_price	float64
discounted_price	float65
image	object
is_FK_Advantage_product	bool
description	object
product_rating	object
overall_rating	object
brand	object
product_specifications	object
product_category_tree	object

Exploratory Data Analysis

The e-commerce data set has 15 attributes out of these labels, and we need the product name and description for this project.

Let's import all the libraries required.

```
#Data Manipulation
import pandas as pd
import numpy as np

# Visualization
import matplotlib.pyplot as plt
import seaborn as sns
```

```
#NLP for text pre-processing
import nltk
import scipy
import re
from scipy import spatial
from nltk.tokenize.toktok import ToktokTokenizer
from nltk.corpus import stopwords
from nltk.tokenize import sent_tokenize, word_tokenize
from nltk.stem import PorterStemmer
tokenizer = ToktokTokenizer()

# other libraries
import gensim
from gensim.models import Word2Vec
import itertools
from sklearn.feature_extraction.text import TfidfVectorizer
from sklearn.decomposition import PCA

# Import linear_kernel
from sklearn.metrics.pairwise import linear_kernel

# remove warnings
import warnings
warnings.filterwarnings(action = 'ignore')
```

Let's load the data set which you downloaded and saved in your local (see Figure 4-5).

	uniq_id	crawl_timestamp	product_url	product_name	product_category_tree	pid r
0	c2d766ca982eca8304150849735ffef9	2016-03-25 22:59:23 +0000	http://www.flipkart.com/alisha-solid-women-s-c...	Alisha Solid Women's Cycling Shorts	["Clothing >> Women's Clothing >> Lingerie, Sl...	SRTEH2FF9KEDEFGF
1	7f7036a6d550aaa89d34c77bd39a5e48	2016-03-25 22:59:23 +0000	http://www.flipkart.com/fabhomedecor-fabric-do...	FabHomeDecor Fabric Double Sofa Bed	["Furniture >> Living Room Furniture >> Sofa B...	SBEEH3QGU7MFYJFY
2	f449ec65dcbc041b6ae5e6a32717d01b	2016-03-25 22:59:23 +0000	http://www.flipkart.com/aw-bellies/p/itmeh4grg...	AW Bellies	["Footwear >> Women's Footwear >> Ballerinas >...	SHOEH4GRSUBJGZXE
3	0973b37acd0c864e3de26e97e5571454	2016-03-25 22:59:23 +0000	http://www.flipkart.com/alisha-solid-women-s-c...	Alisha Solid Women's Cycling Shorts	["Clothing >> Women's Clothing >> Lingerie, Sl...	SRTEH2F6HUZMQ6SJ
4	bc940ea42ee6bef5ac7cea3fb5cfbee7	2016-03-25 22:59:23 +0000	http://www.flipkart.com/sicons-all-purpose-arn...	Sicons All Purpose Arnica Dog Shampoo	["Pet Supplies >> Grooming >> Skin & Coat Care...	PSOEH3ZYDMSYARJ5

Figure 4-5. *Sample data set*

```
data=pd.read_csv("flipkart_com-ecommerce_sample1.csv")
data.head()
```

```
data.shape
```

```
(20000, 15)
```

```
data.info()
```

```
<class 'pandas.core.frame.DataFrame'>
RangeIndex: 19998 entries, 0 to 19997
Data columns (total 15 columns):
uniq_id                     19998 non-null object
crawl_timestamp             19998 non-null object
product_url                 19998 non-null object
product_name                19998 non-null object
product_category_tree       19998 non-null object
pid                         19998 non-null object
retail_price                19920 non-null float64
discounted_price            19920 non-null float64
image                       19995 non-null object
is_FK_Advantage_product     19998 non-null bool
description                 19998 non-null object
product_rating              19998 non-null object
overall_rating              19998 non-null object
brand                       14135 non-null object
product_specifications      19984 non-null object
dtypes: bool(1), float64(2), object(12)
memory usage: 1.2+ MB
```

Here are the observations.

- The data set has a total of 15 columns and 20,000 observations.

- is_FK_Advantage_product is a boolean, the retail_price and discounted_price columns are numerical, and the remaining are categorical.

97

Let's add a new length column to give the total length of the 'description' input variable.

```
data['length']=data['description'].str.len()
```

Add a new column for the number of words in the description before text preprocessing.

```
data['no_of_words'] = data.description.apply(lambda x : len(x.split()))
```

The following is the word count distribution for 'description'.

```
bins=[0,50,75, np.inf]
data['bins']=pd.cut(data.no_of_words, bins=[0,100,300,500,800, np.inf],
labels=['0-100', '100-200', '200-500','500-800' ,'>800'])
words_distribution = data.groupby('bins').size().reset_index().
rename(columns={0:'word_counts'})
sns.barplot(x='bins', y='word_counts', data=words_distribution).
set_title("Word distribution per bin")
```

Figure 4-6 shows that most descriptions have less than 100 words and 20% have 100 to 200 words.

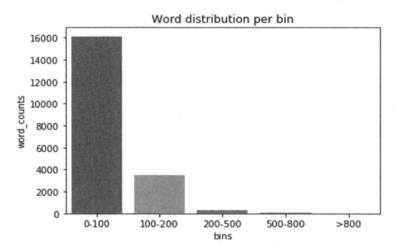

Figure 4-6. *Word distribution of description column*

Now, let's do some data preprocessing.

Data Preprocessing

Data preprocessing includes data cleaning, preparation, transformation, and dimensionality reduction, which convert the raw data into a form that is suitable for further processing.

```
# Number of missing values in each column
missing = pd.DataFrame(data.isnull().sum()).rename(columns = {0:
'missing'})

# Create a percentage of missing values
missing['percent'] = missing['missing'] / len(data)

# sorting the values in desending order to see highest count on the top
missing.sort_values('percent', ascending = False)
```

Figure 4-7 shows that nearly 30% of the brand variables have missing values. Other variables have a negligible number of missing values.

	missing	percent
brand	5863	0.293179
retail_price	78	0.003900
discounted_price	78	0.003900
product_specifications	14	0.000700
image	3	0.000150
description	0	0.000000
no_of_words	0	0.000000
overall_rating	0	0.000000
product_rating	0	0.000000
uniq_id	0	0.000000
is_FK_Advantage_product	0	0.000000
crawl_timestamp	0	0.000000
pid	0	0.000000
product_category_tree	0	0.000000
product_name	0	0.000000
product_url	0	0.000000
bins	0	0.000000

Figure 4-7. *Missing value distribution*

Again, let's move into text preprocessing using multiple methods.

Text Preprocessing

There is a lot of unwanted information present in the text data. Let's clean it up.

Text preprocessing tasks include

- Converting the text data to lowercase

- Removing/replacing the punctuations

- Removing/replacing the numbers

- Removing extra whitespaces

- Removing stop words

- Stemming and lemmatization

```python
# Remove punctuation
data['description'] = data['description'].str.replace(r'[^\w\d\s]', ' ')

# Replace whitespace between terms with a single space
data['description'] = data['description'].str.replace(r'\s+', ' ')

# Remove leading and trailing whitespace
data['description'] = data['description'].str.replace(r'^\s+|\s+?$', '')

# converting to lower case
data['description'] = data['description'].str.lower()

data['description'].head()
```

```
0    key features of alisha solid women s cycling s...
1    fabhomedecor fabric double sofa bed finish col...
2    key features of aw bellies sandals wedges heel...
3    key features of alisha solid women s cycling s...
4    specifications of sicons all purpose arnica do...
Name: description, dtype: object
```

```python
# Removing Stop words
stop = stopwords.words('english')
```

```
pattern = r'\b(?:{})\b'.format('|'.join(stop))
data['description'] = data['description'].str.replace(pattern, '')

# Removing single characters
data['description'] = data['description'].str.replace(r'\s+', ' ')
data['description'] = data['description'].apply(lambda x: " ".join(x for x
in x.split() if len(x)>1))

# Removing domain related stop words from description
specific_stop_words = [ "rs","flipkart","buy","com","free","day","cash","re
placement","guarantee","genuine","key","feature","delivery","products","pro
duct","shipping", "online","india","shop"]
data['description'] = data['description'].apply(lambda x: " ".join(x for x
in x.split() if x not in specific_stop_words))

data['description'].head()

0    features alisha solid women cycling shorts cot...
1    fabhomedecor fabric double sofa bed finish col...
2    features aw bellies sandals wedges heel casual...
3    features alisha solid women cycling shorts cot...
4    specifications sicons purpose arnica dog shamp...
Name: description, dtype: object
```

Let's also see what are the most occurred words in the corpus and understand the data better.

```
#Top frequent words after removing domain related stop words
a = data['description'].str.cat(sep=' ')
words = nltk.tokenize.word_tokenize(a)
word_dist = nltk.FreqDist(words)
word_dist.plot(10,cumulative=False)
print(word_dist.most_common(10))
```

Figure 4-8 shows that data has words like *women, price,* and *shirt* appeared commonly in the data because there are a lot of fashion-related items and most of it is for women.

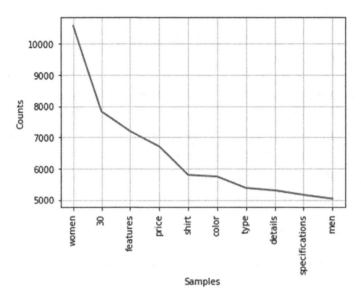

Figure 4-8. *Top frequent words*

Model Building

So far, we have tried to understand data to build better solutions. Now we need to solve problems using algorithms.

There are two models we want to build.

- A content-based recommendation system

- A product search engine

Let's use different NLP techniques, such as TF-IDF and word embeddings. TF-IDF and word embeddings can be used with both models. From the implementation point of view, both models are almost the same. But the problem each solves is different.

So, let's use the TF-IDF approach to solve with a content-based recommendation system and word embeddings for the search engine. But note that reserve can also be done.

Let's start with the recommendation system.

Content-based Recommendation System

Now that you know about content-based recommender systems, let's start implementing one.

For content-based systems, let's use the TF-IDF approach.

```
#text cleaning
data['description'] = data['description'].fillna('')

#define the vectorizer
T_vec =  TfidfVectorizer(stop_words='english')

# get the vectors
T_vec_matrix = T_vec.fit_transform(data['description'])

#shape
T_vec_matrix.shape

(19998, 26162)
```

There are 26,000 unique words in the description.

Next, let's calculate similarity scores for each combination and generate matrix.

A cosine similarity is used in this project. We need to write a function that takes product descriptions as input and lists N most similar items/products.

We also need to do reverse mapping of product names to their indices.

```
# Reversing the map of indices and product

product_index = pd.Series(data.index, index=data['product_name']).drop_
duplicates()
product_index
```

Figure 4-9 shows the output.

```
product_name
Alisha Solid Women's Cycling Shorts                                        0
FabHomeDecor Fabric Double Sofa Bed                                        1
AW Bellies                                                                 2
Alisha Solid Women's Cycling Shorts                                        3
Sicons All Purpose Arnica Dog Shampoo                                      4
Eternal Gandhi Super Series Crystal Paper Weights  with Silver Finish      5
Alisha Solid Women's Cycling Shorts                                        6
FabHomeDecor Fabric Double Sofa Bed                                        7
dilli bazaaar Bellies, Corporate Casuals, Casuals                          8
Alisha Solid Women's Cycling Shorts                                        9
Ladela Bellies                                                            10
Carrel Printed Women's                                                    11
Sicons All Purpose Tea Tree Dog Shampoo                                   12
```

Figure 4-9. *Product names with index*

In the following steps, everything is wrapped under a single function to make testing easier.

1. Obtain the index given the product.

2. Obtain cosine similarity scores.

3. Sort the scores.

4. Get the top N results from the list.

5. Output the product names.

```python
# Function that takes in product title as input and outputs the most
similar product

def predict_products(text):

    # getting index
    index = product_index[text]

    # Obtaining the pairwsie similarity scores
    score_matrix = linear_kernel(T_vec_matrix[index], T_vec_matrix)
    matching_sc= list(enumerate(score_matrix[0]))

    # Sort the product based on the similarity scores
    matching_sc= sorted(matching_sc, key=lambda x: x[1], reverse=True)
```

```
    # Getting the scores of the 10 most similar product
    matching_sc= matching_sc[1:10]

    # Getting the product indices
    product_indices = [i[0] for i in matching_sc]

    # Show the similar products
    return data['product_name'].iloc[product_indices]

recommended_product = predict_products(input("Enter a product name: "))
if recommended_product is not None:
    print ("Similar products")
    print("\n")
    for product_name in recommended_product:
        print (product_name)
```

```
Enter a product name:  Engage Urge and Urge Combo Set
```

```
Enter a product name: Engage Urge and Urge Combo Set
Similar products

Engage Rush and Urge Combo Set
Engage Urge-Mate Combo Set
Engage Jump and Urge Combo Set
Engage Fuzz and Urge Combo Set
Engage Mate+Urge Combo Set
Engage Urge+Tease Combo Set
Engage Combo Set
Engage Combo Set
Engage Combo Set
```

Let's look at one more example.

```
Enter a product name:  Lee Parke Running Shoes
```

```
Enter a product name: Lee Parke Running Shoes
Similar products
```

105

```
Lee Parke Walking Shoes
N Five Running Shoes
Knight Ace Kraasa Sports Running Shoes, Cycling Shoes, Walking Shoes
WorldWearFootwear Running Shoes, Walking Shoes
reenak Running Shoes
Chazer Running Shoes
Glacier Running Shoes
Sonaxo Men Running Shoes
ETHICS Running Shoes
```

Observe the results. If a customer clicks Lee Parke Running Shoes, they get recommendations based on any other brand running shoes or Lee Parke's any other products.

- Lee Parke Walking Shoes is there because of the Lee Parke brand.

- The rest of the recommendations are running shoes by a different brand.

You can also add price as a feature and get only products in the price range of the customer's selected product.

This is one version of the recommendation system using NLP. To get better results, you can do the following things.

- A better approach to the content-based recommender system can be applied by creating the user profile (currently not in the scope of the data set).

- Use word embeddings as features.

- Try different distance measures.

That's it. We explored how to build a recommendation system using natural language processing.

Now let's move on to another interesting use case which is a product search engine.

Product Search Engine

The next problem statement is optimizing the search engine to get better search results. The biggest challenge is that most search engines are string matching and might not perform well in all circumstances.

For example, if the user searches "guy shirt", the search results should have all the results that have men, a boy, and so on. The search should not work based on string match, but the other similar words should also consider.

The best way to solve this problem is *word embeddings*.

Word embeddings are N-dimensional vectors for each word that captures the meaning of the word along with context.

word2vec is one of the methods to construct such an embedding. It uses a shallow neural network to build the embeddings. There are two ways the embeddings can be built: skip-gram and CBOW (common bag-of-words).

The CBOW method takes each word's context as the input and predicts the word corresponding to the context. The input to the network context and passed to the hidden layer with N neurons. Then at the end, the output layer predicts the word using the softmax layer. The hidden layer neuron's weight is considered the vector that captured the meaning and context.

The skip-gram model is the reverse of CBOW. The word is the input, and the network predicts context.

That's the brief theory about word embeddings and how it works. We can build the embeddings or use existing trained ones. It takes a lot of data and resources to build one, and for domains like healthcare, we need to build our own because generalized embeddings won't perform well.

Implementation

Let's use the pretrained word2vec model on the news data set by Google. The trained model can be imported, and vectors can be obtained for each word. Then, any of the similarity measures can be leveraged to rank the results.

```
#Creating list containing description of each product as sublist
fin=[]
for i in range(len(data['description'])):
    temp=[]
    temp.append(data['description'][i])
```

```python
    fin = fin + temp

data1 = data[['product_name','description']]

#import the word2vec

from gensim.models import KeyedVectors
filename = 'C:\\GoogleNews-vectors-negative300.bin'
model = KeyedVectors.load_word2vec_format(filename, binary=True,
limit=50000)

#Preprocessing

def remove_stopwords(text, is_lower_case=False):
    pattern = r'[^a-zA-z0-9\s]'
    text = re.sub(pattern, '', text[0])
    tokens = nltk.word_tokenize(text)
    tokens = [token.strip() for token in tokens]
    if is_lower_case:
        filtered_tokens = [token for token in tokens if token not in stop]
    else:
        filtered_tokens = [token for token in tokens if token.lower() not
        in stop]
    filtered_text = ' '.join(filtered_tokens)
    return filtered_text

# Obtain the embeddings, lets use "300"

def get_embedding(word):
    if word in model.wv.vocab:
        return model[word]
    else:
        return np.zeros(300)
```

For every document, let's take the mean of all the words present in the document.

```
# Obtaining the average vector for all the documents

out_dict =  {}
for sen in fin:
    average_vector = (np.mean(np.array([get_embedding(x) for x in nltk.
    word_tokenize(remove_stopwords(sen))]), axis=0))
    dict = { sen : (average_vector) }
    out_dict.update(dict)

# Get the similarity between the query and documents

def get_sim(query_embedding, average_vector_doc):
    sim = [(1 - scipy.spatial.distance.cosine(query_embedding, average_
    vector_doc))]
    return sim

# Rank all the documents based on the similarity

def Ranked_documents(query):
    global rank
    query_words =  (np.mean(np.array([get_embedding(x) for x in nltk.word_
    tokenize(query.lower())],dtype=float), axis=0))
    rank = []
    for k,v in out_dict.items():
        rank.append((k, get_sim(query_words, v)))
    rank = sorted(rank,key=lambda t: t[1], reverse=True)
    dd =pd.DataFrame(rank,columns=['Desc','score'])
    rankfin = pd.merge(data1,dd,left_on='description',right_on='Desc')
    rankfin = rankfin[['product_name','description','score']]
    print('Ranked Documents :')
    return rankfin

# Call the IR function with a query
query=input("What would you like to search")
Ranked_documents(query)

# output
```

```
What would you like to searchbag
Ranked Documents :
```

	product_name	description	score
0	Alisha Solid Women's Cycling Shorts	key features alisha solid women cycling shorts...	[1.0000000515865854]
1	FabHomeDecor Fabric Double Sofa Bed	fabhomedecor fabric double sofa bed finish col...	[1.0000000515865854]
2	AW Bellies	key features aw bellies sandals wedges heel ca...	[1.0000000515865854]
3	Alisha Solid Women's Cycling Shorts	key features alisha solid women cycling shorts...	[1.0000000515865854]

Figure 4-10. *Model output*

Advanced Search Engine Using PyTerrier and Sentence-BERT

Let's few advanced deep learning-based solutions to solve this problem. Figure 4-11 shows the entire framework for this approach.

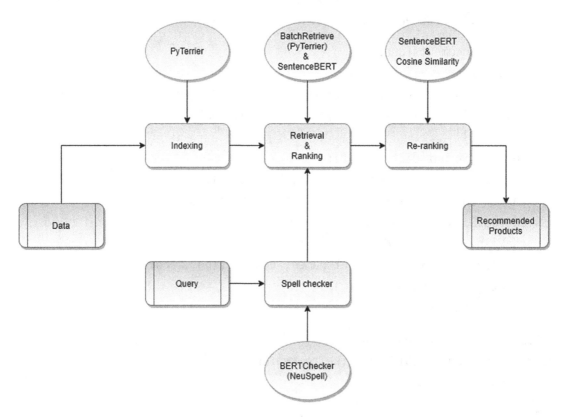

Figure 4-11. *Pipeline of the implementation of the PyTerrier-based search engine*

Indexing is an important part of information retrieval (IR) systems. For indexing, we use DFIndexer. Indexing simplifies the retrieval process.

BatchRetrieve is one of the most widely used PyTerrier objects. It uses a pre-existing Terrier index data structure.

NeuSpell is an open source package for correcting spellings based on the context. This package has ten spell-checkers based on various neural models. To implement this model, import the BERTChecker package from NeuSpell.

BERTChecker works for multiple languages, including English, Arabic, Hindi, and Japanese.

Let's use the PyTerrier and Sentence-BERT libraries and proceed with implementation.

The following installs the required packages and libraries.

```python
import pandas as pd
import numpy as np
import seaborn as sns
import matplotlib.pyplot as plt
import string
import re
%matplotlib inline
import nltk
nltk.download('punkt')
nltk.download('wordnet')
nltk.download('stopwords')
from nltk.corpus import stopwords
from nltk.tokenize import word_tokenize
from nltk.stem.wordnet import WordNetLemmatizer
lem = WordNetLemmatizer()
stop_words = set(stopwords.words('english'))
exclude = set(string.punctuation)
import string

!pip install python-terrier
import pyterrier as pt
if not pt.started():
  pt.init()
```

```
!pip install -U sentence-transformers

!pip install neuspell
!pip install -e neuspell/
!git clone https://github.com/neuspell/neuspell; cd neuspell
import os
os.chdir("/content/neuspell")

!pip install -r /content/neuspell/extras-requirements.txt
!python -m spacy download en_core_web_sm

#Unzipping the multi-linguistic packages
!wget https://storage.googleapis.com/bert_models/2018_11_23/multi_cased_
L-12_H-768_A-12.zip
!unzip *.zip

#importing neuspell
from neuspell import BertChecker
from sklearn.metrics.pairwise import cosine_similarity
from sentence_transformers import SentenceTransformer
model = SentenceTransformer('sentence-transformers/bert-base-nli-mean-
tokens')
```

Load the data set.

```
df=pd.read_csv(flipkart_com-ecommerce_sample.csv)
df.head()
```

Data Preprocessing

Let's do some more text preprocessing.

First, make the 'category_tree' column a simple list.

```
df['product_category_tree']=df['product_category_tree'].map(lambda x:x.
strip('[]'))
df['product_category_tree']=df['product_category_tree'].map(lambda x:x.
strip('"'))
df['product_category_tree']=df['product_category_tree'].map(lambda x:x.
split('>>'))
```

Next, drop unwanted columns.

```
df.drop(['crawl_timestamp','product_url','image',"retail_
price","discounted_price","is_FK_Advantage_product","product_
rating","overall_rating","product_specifications"],axis=1,inplace=True)
```

Then, drop duplicate products.

```
uniq_prod=df.copy()
uniq_prod.drop_duplicates(subset ="product_name", keep = "first", inplace =
True)
```

Remove stop words and punctuations and then perform tokenization and lemmatization.

```
def filter_keywords(doc):
    doc=doc.lower()
    stop_free = " ".join([i for i in doc.split() if i not in stop_words])
    punc_free = "".join(ch for ch in stop_free if ch not in exclude)
    word_tokens = word_tokenize(punc_free)
    filtered_sentence = [(lem.lemmatize(w, "v")) for w in word_tokens]
    return filtered_sentence
```

Apply the filter_keywords function to selected columns to obtain the keywords for each column.

```
uniq_prod['product'] = uniq_prod['product_name'].apply(filter_keywords)
uniq_prod['description'] = uniq_prod['description'].astype("str").
apply(filter_keywords)
uniq_prod['brand'] = uniq_prod['brand'].astype("str").apply(filter_
keywords)
```

Combine all the keywords for each product.

```
uniq_prod["keywords"]=uniq_prod['product']+uniq_prod['brand']+ df['product_
category_tree']+uniq_prod['description']
uniq_prod["keywords"] = uniq_prod["keywords"].apply(lambda x: ' '.join(x))
```

Creating a 'docno' column, which gives recommendations.

```
uniq_prod['docno']=uniq_prod['product_name']
```

Drop unwanted columns.

```
uniq_prod.drop(['product','brand','pid','product_
name'],axis=1,inplace=True)
```

```
uniq_prod.head()
```

Figure 4-12 shows the final data set that we are using going forward. We perform the query search in 'keywords' and obtain the corresponding products based on the most relevant keywords.

	uniq_id	product_category_tree	description	keywords	docno
0	c2d766ca982eca8304150849735ffef9	[Clothing , Women's Clothing , Lingerie, Sle...	[key, feature, alisha, solid, womens, cycle, s...	alisha solid womens cycle short alisha Clothin...	Alisha Solid Women's Cycling Shorts
1	7f7036a6d550aaa89d34c77bd39a5e48	[Furniture , Living Room Furniture , Sofa Be...	[fabhomedecor, fabric, double, sofa, bed, fini...	fabhomedecor fabric double sofa bed fabhomedec...	FabHomeDecor Fabric Double Sofa Bed
2	f449ec65dcbc041b6ae5e6a32717d01b	[Footwear , Women's Footwear , Ballerinas , ...	[key, feature, aw, belly, sandals, wedge, heel...	aw belly aw Footwear Women's Footwear Ball...	AW Bellies
4	bc940ea42ee6bef5ac7cea3fb5cfbee7	[Pet Supplies , Grooming , Skin & Coat Care ...	[specifications, sicons, purpose, arnica, dog,...	sicons purpose arnica dog shampoo sicons Pet S...	Sicons All Purpose Arnica Dog Shampoo
5	c2a17313954882c1dba461863e98adf2	[Eternal Gandhi Super Series Crystal Paper Wei...	[key, feature, eternal, gandhi, super, series,...	eternal gandhi super series crystal paper welg...	Eternal Gandhi Super Series Crystal Paper Weig...

Figure 4-12. Data set snapshot

Building the Search Engine

Let's use the DFIndexer object to create the index for keywords.

```
!rm -rf /content/iter_index_porter
pd_indexer = pt.DFIndexer("/content/pd_index")
indexref = pd_indexer.index(uniq_prod["keywords"], uniq_prod["docno"])
```

Let's implementing the NeuSpell spell checker on the user query and save it to an object.

```
spellcheck = BertChecker()
spellcheck.from_pretrained(
    ckpt_path=f"/content/multi_cased_L-12_H-768_A-12"  # "<folder where the
    model is saved>"
)
```

```
X=input("Search Engine:")
query=spellcheck.correct(X)
print(query
```

```
Search Engine:womns clothing
women clothing
```

Perform ranking and retrieval using PyTerrier and Sentence-BERT.

```
prod_ret = pt.BatchRetrieve(indexref, wmodel='TF_IDF',
properties={'termpipelines': 'Stopwords'})
pr=prod_ret.compile()
output=pr.search(query)
docno=list(output['docno'])
transform=model.encode(docno)
```

Create embeddings and re-ranking using PyTerrier and cosine similarity.

```
embedding={}
for i,product in enumerate(docno):
  embedding[product]=transform[i]
```

```
q_embedding=model.encode(query).reshape(1,-1)
l=[]
for product in embedding.keys():
score=cosine_similarity(q_embedding,embedding[product].reshape(1,-1))[0][0]
l.append([product,score])
```

```
output2=pd.DataFrame(l,columns=['product_name','score'])
```

Let's look at the results.

```
output2.sort_values(by='score',ascending=False).head(10)
```

Figure 4-13 shows "women clothing" as the query. In the output, there is a list of products that are a part of women's clothing. The corresponding scores represent relevance. Note that the results are very relevant to the search query.

	product_name	score
95	People Women's Dress	0.862905
212	Nasha Women's Gathered Dress	0.763813
166	Kasturi Women's Gathered Dress	0.754854
207	Sbuys Women's Gathered Dress	0.751761
111	Teemoods Women's Camisole	0.748863
55	Viyasha Women's, Girl's Shapewear	0.747937
182	Karishma Women's Gathered Dress	0.742156
225	Kwardrobe Women's Gathered Dress	0.741498
200	Modimania Women's Gathered Dress	0.740214
722	IDK Woman Women's Shift Dress	0.736790

Figure 4-13. *Model output*

Multilingual Search Engine Using Deep Text Search

One of the challenging tasks in these search engines is the languages. These products are very regional, and English is not the only language spoken. To solve this, we can use Deep Text Search.

Deep Text Search is an AI-based multilingual text search engine with transformers. It supports 50+ languages. The following are some of its features.

- It has a faster search capability.

- It has very accurate recommendations.

- It works best for implementing Python-based applications.

Let's use the following data sets to understand this library.

- The English data set contains 30 rows and 13 columns.

https://data.world/login?next=%2Fpromptcloud%2Fwalmart-
product-data-from-usa%2Fworkspace%2Ffile%3Ffilename%3Dwa
lmart_com-ecommerce_product_details__20190311_20191001_
sample.csv

- The Arabic data set is a text file related to the Arabic newspaper corpus.

  https://www.kaggle.com/abedkhooli/arabic-bert-corpus

- The Hindi data set contains 900 movie reviews in three classes (positive, neutral, negative) collected from Hindi news websites.

  https://www.kaggle.com/disisbig/hindi-movie-reviews-
  dataset?select=train.csv

- The Japanese data set contains the Japanese prime minister's tweets.

  https://www.kaggle.com/team-ai/shinzo-abe-japanese-
  prime-minister-twitter-nlp

Let's start with the English data set.

Install the required packages and importing libraries.

```
!pip install neuspell
!pip install -e neuspell/
!git clone https://github.com/neuspell/neuspell; cd neuspell
!pip install DeepTextSearch

import os
os.chdir("/content/neuspell")

!pip install -r /content/neuspell/extras-requirements.txt
!python -m spacy download en_core_web_sm

#Unzipping the multi-linguistic packages
!wget https://storage.googleapis.com/bert_models/2018_11_23/multi_cased_L-
12_H-768_A-12.zip
!unzip *.zip

# importing nltk
import nltk
```

```
nltk.download('punkt')
nltk.download('averaged_perceptron_tagger')
nltk.download('maxent_ne_chunker')
nltk.download('words')
nltk.download('wordnet')

#import DeepTextSearch
from DeepTextSearch import TextEmbedder,TextSearch,LoadData
from nltk.corpus import wordnet
import pandas as pd
from neuspell import BertChecker
```

Let's use BERTChecker for spell checking. It also supports multiple languages.

```
spellcheck = BertChecker()
spellcheck.from_pretrained(
    ckpt_path=f"/content/multi_cased_L-12_H-768_A-12")
# "<folder where the model is saved>"
```

Let's input the query and check how its works.

```
X=input("Enter Product Name:")
y=spellcheck.correct(X)
print(y)

Enter Product Name: shirts
shirts
```

Let's also use POS tagging to select relevant words from the given query.

```
#function to get the POS tag
def preprocess(sent):
    sent = nltk.word_tokenize(sent)
    sent = nltk.pos_tag(sent)
    return sent

sent = preprocess(y)
l=[]
for i in sent:
  if i[1]=='NNS' or i[1]=='NN':
```

```
    l.append(i[0])
print(l)
```

```
['shirts']
```

In the next step, let's use query expansion to get synonyms of words, so that we can get more relevant recommendations.

```
query=""
for i in l:
  query+=i
  synset = wordnet.synsets(i)
  query+=" "+synset[0].lemmas()[0].name()+" "

print(query)
```

```
shirts shirt
```

We created this dictionary to display the product names as per the recommendations given in the description.

```
#importing the data
df=pd.read_csv("walmart_com-ecommerce_product_details__20190311_20191001_
sample.csv")

df1=df.set_index("Description", inplace = False)

df2=df1.to_dict()
dict1=df2['Product Name']
```

Embed the data in a pickle file as the library requires it to be in that format.

```
data = LoadData().from_csv("walmart_com-ecommerce_product_
details__20190311_20191001_sample.csv")
TextEmbedder().embed(corpus_list=data)
corpus_embedding = TextEmbedder().load_embedding()
```

Search the ten most relevant products based on the query.

```
n=10
t=TextSearch().find_similar(query_text=query,top_n=n)
```

```
for i in range(n):
  t[i]['text']=dict1[t[i]['text']]
  print(t[i])
```

Figure 4-14 shows the results for the "shirts" search query.

```
{'index': 19, 'text': "Men's Big & Tall Harbor Bay Space-Dye Piqué Polo Shirt", 'score': 0.34476066}
{'index': 18, 'text': "Men's Big & Tall Harbor Bay Space-Dye Piqué Polo Shirt", 'score': 0.34476066}
{'index': 17, 'text': "Men's Big & Tall Harbor Bay Space-Dye Piqué Polo Shirt", 'score': 0.34476066}
{'index': 16, 'text': "Men's Big & Tall Harbor Bay Space-Dye Piqué Polo Shirt", 'score': 0.34476066}
{'index': 15, 'text': "Men's Big & Tall Harbor Bay Space-Dye Piqué Polo Shirt", 'score': 0.34476066}
{'index': 13, 'text': "Men's Big & Tall Harbor Bay Space-Dye Piqué Polo Shirt", 'score': 0.34476066}
{'index': 12, 'text': "Men's Big & Tall Harbor Bay Space-Dye Piqué Polo Shirt", 'score': 0.34476066}
{'index': 11, 'text': "Men's Big & Tall Harbor Bay Space-Dye Piqué Polo Shirt", 'score': 0.34476066}
{'index': 20, 'text': "Men's Big & Tall Harbor Bay Space-Dye Piqué Polo Shirt", 'score': 0.34476066}
{'index': 10, 'text': "Men's Big & Tall Harbor Bay Space-Dye Piqué Polo Shirt", 'score': 0.34476066}
```

Figure 4-14. *Model output*

Now let's see how the search works in Arabic.

```
X1=input("Search Engine:")
y1=spellcheck.correct(X1)
print(y1)
```

Search Engine:صاحب
صاحب

Let's import the Arabic data corpus to perform the search.

```
# import library and data
from DeepTextSearch import LoadData

data1 = LoadData().from_text("wiki_books_test_1.txt")
TextEmbedder().embed(corpus_list=data1)
corpus_embedding = TextEmbedder().load_embedding()
```

```
Embedding data Saved Successfully Again!
['corpus_embeddings_data.pickle', 'corpus_list_data.pickle']
Embedding data Loaded Successfully!
['corpus_embeddings_data.pickle', 'corpus_list_data.pickle']
```

Let's find the top 10 relevant documents using the textseach function. Figure 4-15 shows the output.

```
TextSearch().find_similar(query_text=y1,top_n=10)
```

```
[{'index': 383, 'score': 0.79338586, 'text': 'وفيهما تجب :والثاني.'},
 {'index': 273, 'score': 0.74463975, 'text': 'بالثاني :والثاني.'},
 {'index': 433, 'score': 0.7429859, 'text': 'القولين على :والثاني.'},
 {'index': 32,
  'score': 0.7291739,
  'text': 'أعلم والله الاكثرون وصححه ،أصح الثاني :قلت.'},
 {'index': 242, 'score': 0.7224536, 'text': 'بعذر :الثاني وعلى.'},
 {'index': 460, 'score': 0.71352804, 'text': 'استحبابه :والثاني.'},
 {'index': 446, 'score': 0.71352804, 'text': 'استحبابه :والثاني.'},
 {'index': 8885, 'score': 0.7128246, 'text': 'الثاني بالقسم وجه ومن . .'},
 {'index': 205, 'score': 0.68822443, 'text': 'لا :والثاني.'},
 {'index': 4641, 'score': 0.68177044, 'text': 'ثالث خبر ولواحق.'}]
```

Figure 4-15. *Model output*

Now let's see how the search works in Hindi.

```
X_hindi=input("Search Engine:")
y_hindi=spellcheck.correct(X_hindi)
print(y_hindi)
```

```
Search Engine:निर्देशक
निर्देशक
```

```
#loading the Hindi data corpus
data_hindi = LoadData().from_csv("hindi.csv")
TextEmbedder().embed(corpus_list=data_hindi)
corpus_embedding = TextEmbedder().load_embedding()
```

Let's find the top 10 relevant results. Figure 4-16 show the output.

```
TextSearch().find_similar(query_text=y_hindi,top_n=10)
```

```
[{'index': 410,
  'score': 0.67329043,
  'text': 'बैनर :\nयूटीवी मोशन पिक्वर्स\n\nनिर्माता :\nरॉनी स्क्रूवाला\n\nनिर्देशन एवं संगीत :\nविशाल भारद्वाज\n\nगीत :\nगुलजार\n\nकलाकार :\nशाहिद कपूर, प्रियंका चोपड़ा, अमृ
 {'index': 651,
  'score': 0.6686705,
  'text': 'बैनर :\nहरी ओम एंटरटेनमेंट कं., ध्रीज़ कंपनी, यूटीवी मोशन पिक्चर्स\n\nनिर्माता :\nफराह खान, अक्षय कुमार, शिरीष कुंदर\n\nनिर्देशक :\nशिरीष कुंदर\n\nसंगीत :\nगौर
 {'index': 463,
  'score': 0.66178524,
  'text': '\n\nश्री-ड\nok\nबैनर :\nयूटीवी स्पॉट बॉय\n\nनिर्माता :\nसिद्धार्थ रॉय कपूर, रॉनी स्क्रूवाला\n\nनिर्देशक :\nरेमो डिसूजा\n\nसंगीत :\nसचिन जिगर\n\nकलाकार :\nप्रभुदेवा,
 {'index': 39,
  'score': 0.6527176,
  'text': 'बैनर :\nभंडारकर एंटरटेनमेंट, वाइड फ्रेम पिक्वर्स\n\nनिर्माता :\nकुमार मंगत पाठक, मधुर भंडारकर\n\nनिर्देशक :\nमधुर भंडारकर\n\nसंगीत :\nप्रीतम चक्रवर्ती\n\nकलाका
 {'index': 23,
  'score': 0.6512797,
  'text': 'बैनर :\nपीवीआर पिक्वर्स\n\nनिर्माता :\nअजय बिजली, दिबाकर बैनर्जी, प्रिया श्रीधरन, संजीव के. बिजली\n\nनिर्देशक :\nदिबाकर बैनर्जी\n\nसंगीत :\nविशाल-शेखर\n\nकल
 {'index': 359,
  'score': 0.6375189,
  'text': 'कुल मिलाकर कहा जा सकता है कि 'हरि पुत्तर' समय और पैसे की बर्बादी है।\nनिर्माता :\nलकी कोहली, मुनीष पुरी, ए पी पारिंगी\n\nनिर्देशक :\nलकी कोहली, राजेश बज
 {'index': 84,
  'score': 0.63575315,
  'text': 'बैनर :\nसरोज एंटरटेनमेंट प्रा.लि.\n\nनिर्माता :\nरचना सुनील सिंह\n\nनिर्देशक :\nपार्थ घोष\n\nकलाकार :\nजैकी श्रॉफ, मनीषा कोइराला, निकिता आनंद, रोज़ा\n\n\n\n\n
 {'index': 151,
  'score': 0.62818766,
  'text': 'बैनर :\nवाय फिल्म्स\n\nनिर्माता :\nआशीष पाटिल\n\nनिर्देशक :\nबम्मी\n\nसंगीत :\nराम सम्पत\n\nकलाकार :\nश्रद्धा कपूर, ताहा शाह, शेनाज ट्रेज़रीवाला, जन्नत ज़ुबैर र
 {'index': 121,
  'score': 0.62714005,
  'text': 'बैनर :\nयशराज फिल्म्स\n\nनिर्माता :\nआदित्य चोपड़ा\n\nनिर्देशक :\nमनीष शर्मा\n\nसंगीत :\nसलीम मर्चेंट-सुलेमान मर्चेंट\n\nकलाकार :\nरणवीर सिंह, अनुष्का शर्मा, परि
 {'index': 10,
  'score': 0.62641734,
  'text': "IFM \r\t\t\t\t\t\t\t\tIFM \r\t\t\t\t\t\t\t\t \nIFM \r\t\t\t\t\t\t\t\tIFM \r\t\t\t\t\t\t\t\t \nनिर्माता :\nआदित्य चोपड़ा\nनिर्देशक :\nनि
```

Figure 4-16. *Model output*

Let's try one more language, Japanese.

```
X_japanese=input("Search Engine:")
# y_japanese=spellcheck.correct(X_japanese)
print(X_japanese)
```

```
Search Engine:経済的
経済的
```

```
#loading the data
data_japanese = LoadData().from_text("Japanes_Shinzo Abe Tweet 20171024 -
Tweet.csv")
TextEmbedder().embed(corpus_list=data_chinese)
```

```
corpus_embedding = TextEmbedder().load_embedding()
```

Find the top ten tweets based on the search query. Figure 4-17 shows the output.

```
TextSearch().find_similar(query_text=X_japanese,top_n=10)
```

```
[{'index': 8,
  'score': 0.37281358,
  'text': 'https://twitter.com/AbeShinzo,安倍晋三,AbeShinzo,Oct 17,https://twitter.com/AbeShinzo/status/920292054663434245,私たち自民党は日本の経済
 {'index': 63,
  'score': 0.36950645,
  'text': 'https://twitter.com/AbeShinzo,安倍晋三,AbeShinzo,31 Mar 2016,https://twitter.com/AbeShinzo/status/715497509011861504,元FRB議長、元財務長
 {'index': 62,
  'score': 0.35948235,
  'text': 'https://twitter.com/AbeShinzo,安倍晋三,AbeShinzo,31 Mar 2016,https://twitter.com/AbeShinzo/status/715497629832970240,Good discussions
 {'index': 14,
  'score': 0.27642867,
  'text': 'https://twitter.com/AbeShinzo,安倍晋三,AbeShinzo,Oct 14,https://twitter.com/AbeShinzo/status/919126975544836096,熊本地震から一年半、被害
 {'index': 21,
  'score': 0.26905608,
  'text': 'https://twitter.com/AbeShinzo,安倍晋三,AbeShinzo,Oct 10,https://twitter.com/AbeShinzo/status/917711860945739776,明日１０月１１日(水) 安倍
 {'index': 1,
  'score': 0.26412064,
  'text': 'https://twitter.com/AbeShinzo,安倍晋三,AbeShinzo,Oct 21,https://twitter.com/AbeShinzo/status/921745765067669505,選挙期間中、自民党の候補
 {'index': 23,
  'score': 0.25998157,
  'text': 'https://twitter.com/AbeShinzo,安倍晋三,AbeShinzo,Oct 9,https://twitter.com/AbeShinzo/status/917368221845434368,明日１０月１０日(火) 安倍
 {'index': 22,
  'score': 0.25498116,
  'text': 'https://twitter.com/AbeShinzo,安倍晋三,AbeShinzo,Oct 10,https://twitter.com/AbeShinzo/status/917680616488943617,いよいよ本日より衆議院総
 {'index': 3,
  'score': 0.2535026,
  'text': 'https://twitter.com/AbeShinzo,安倍晋三,AbeShinzo,Oct 20,https://twitter.com/AbeShinzo/status/921322049573765120,明日１０月２１日(土) 安倍
 {'index': 17,
  'score': 0.2500959,
  'text': 'https://twitter.com/AbeShinzo,安倍晋三,AbeShinzo,Oct 12,https://twitter.com/AbeShinzo/status/918448154474770432,明日１０月１３日(金) 安倍
```

Figure 4-17. *Model output*

Summary

We implemented a search engine and recommendation systems using various models in this chapter. We started with a simple recommender system using the TF-IDF approach to find the similarity score for all the products description. Based on the description, products are ranked and shown to the users. Later, we explored how to build a simple search engine using word embeddings and ranked the results.

Then, we jumped into advanced models like PyTerrier and Sentence-BERT, where pretrained models extract the vectors. Since these models are deep learning-based, results are a lot better when compared to traditional approaches. We also used Deep Text Search, another deep learning library that works for multilanguage text corpus.

## CHAPTER 5

Creating a Résumé Parsing, Screening and Shortlisting System

The objective of this project is to create a résumé shortlisting system using natural language processing.

Context

With millions of individuals looking for a job, it is impossible to sort through hundreds and thousands of résumés to find the best applicant. Most résumés do not follow a standard format; that is, nearly every résumé on the market has a unique structure and content. Human resource personnel must personally review each résumé to determine the best fit for the job description. This is time-consuming and prone to inaccuracy, as a suitable candidate for the job may be overlooked in the process.

One of the difficult challenges companies currently face is choosing the right candidate for a job with a limited amount of time and resources. When choosing the best candidate, a recruiter must consider the following.

- From the set of résumés in the database, manually go through them one by one to look for the those that best match the given job description.

- Out of the selected résumés, rank them based on the greatest relevance to the job description.

- Once ranked, information like name, contact number, and e-mail address, is necessary to take the candidate further.

© Akshay Kulkarni, Adarsha Shivananda and Anoosh Kulkarni 2022
A. Kulkarni et al., *Natural Language Processing Projects*, https://doi.org/10.1007/978-1-4842-7386-9_5

To overcome these challenges, let's see how NLP can build a résumé parsing and shortlisting system that parses, matches, selects, and ranks the résumés from a huge pool based on the job description (JD).

Methodology and Approach

The key requirements to solve this project are job descriptions and résumés that act as input to the model, which is then passed on to the document reader based on the format (word or PDF), which extracts all the content from word/PDF documents in the form of text.

Now that we have the text data, it is important to process it to remove the noise and unwanted information like stop words and punctuation.

Once résumés and job descriptions are processed, we need to convert the text into features using a count vectorizer or TD-IDF. TD-IDF might make more sense for this use case because we want to emphasize the occurrence of each keyword (skill) mentioned in the résumé. After that, we use *truncated singular vector decomposition* to reduce the dimensions of the feature vector.

Then we move on to the model building stage, where we have a similarity engine to capture how similar the two documents (résumé and JD) are with respect to the keywords present in them. Using the similarity scores, we rank each résumé to each job description present in the system. For example, there are two job descriptions (data science and analytics) and 50 résumés from diversified skill sets. So, out of these 50 résumés, this model would provide top résumés for data science and top résumés for analytics.

Once the model is built, and we have selected the top résumés for the given job description, it is important to extract relevant information like the person's name, contact number, location, and e-mail address.

Finally, we collate all the outcomes and present them in a tabular format, which the recruiter can access and proceed with the hiring process. The validation and visualization layer help the recruiter further validate and have a look at the top skills in the résumé without having to open the résumé to understand the candidate's profile.

Figure 5-1 shows the process flow diagram as an approach to solve this problem.

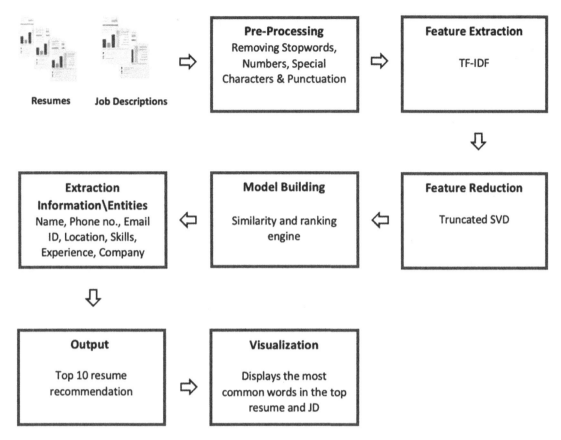

Figure 5-1. *Shows the entire framework for the model*

Note   Extracting of information and entities is done on the original text (without preprocessing).

Implementation

The data set we considered consists of 32 different résumés from various domains and three different job descriptions. This data set is open source. The data sets that are used here contain different file formats of résumés and job descriptions. The following are the links to the data sets. (We considered only a few data sets for this chapter demo, but there are thousands of résumés for each domain profile.)

Now that the data is collected, let's start the implementation.

Installing and Importing Required Libraries

```
# Installing required Libraries
!pip install textract
!pip install -U nltk
!pip install pdfminer3
!pip install mammoth
!pip install locationtagger

# Importing required libraries
import pandas as pd
from google.colab import drive
from pdfminer3.layout import LAParams
from pdfminer3.pdfpage import PDFPage
from pdfminer3.pdfinterp import PDFResourceManager
from pdfminer3.pdfinterp import PDFPageInterpreter
from pdfminer3.converter import TextConverter
import io
import os
import nltk
nltk.download('stopwords')
from nltk.corpus import stopwords
import re
nltk.download('punkt')
nltk.download('averaged_perceptron_tagger')
from sklearn.feature_extraction.text import TfidfVectorizer
from sklearn.metrics.pairwise import cosine_similarity
import en_core_web_sm
nlp = en_core_web_sm.load()
import matplotlib.pyplot as plt
from wordcloud import WordCloud
import mammoth
import locationtagger
```

```
nltk.download('maxent_ne_chunker')
nltk.download('words')
from nltk.corpus import wordnet
nltk.download('wordnet')
from sklearn.decomposition import TruncatedSVD
```

Reading Résumés and Job Descriptions

Now, let's start with creating the path to the directory of both résumés and job descriptions.

```
# Making directory of Resumes and Job Description
directory = '/content/drive/MyDrive/'
resume_path = directory + 'Resumes/'
jd_path = directory + 'JD/'
```

Next, let's extract all the text information from the résumés and job description. Let's write a function that extracts PDF-format résumés. It can also extract the data from tables. Here we use the PDFResourceManager, PDFPageInterpreter, and TextConverter functions, which are defined in pdfminer3.

```
#This function is used to extract text from PDf file. It can also extract
tables from a pdf file
def pdf_extractor(path):
    r_manager = PDFResourceManager()
    output = io.StringIO()
    converter = TextConverter(r_manager, output, laparams=LAParams())
    p_interpreter = PDFPageInterpreter(r_manager, converter)

    with open(path, 'rb') as file:

      for page in PDFPage.get_pages(file,caching=True,check_
      extractable=True):
          p_interpreter.process_page(page)
          text = output.getvalue()
    converter.close()
    output.close()

    return text
```

The following function reads documents for any of the following formats.

- PDF

- DOCX

- DOC

- TXT

```
# A function for reading pdf, docx, doc and txt files
def read_files(file_path):
  fileTXT = []

  # This for loop is for reading all the files in file_path mentioned in
  the function
  for filename in os.listdir(file_path):

    # If the document is in pdf format then this code will be executed
    if(filename.endswith(".pdf")):
        try:
            fileTXT.append(pdf_extractor(file_path+filename)) # Here the
            pdf_extractor function is used to extract pdf file
        except Exception:
            print('Error reading pdf file :' + filename)

    # If the document is in docx format then this code will be executed
    if(filename.endswith(".docx")):
        try:
          with open(file_path + filename, "rb") as docx_file:
            result = mammoth.extract_raw_text(docx_file)
            text = result.value
            fileTXT.append(text)
        except IOError:
          print('Error reading .docx file :')

    # If the given document is in doc format then this loop will be
    executed
    if(filename.endswith(".doc")):
        try:
```

```
            text = textract.process(file_path+filename).decode('utf-8')
            fileTXT.append(text)
        except Exception:
            print('Error reading .doc file :' + filename)

    # If the given file in txt format then this file will be executed
    if(filename.endswith(".txt")):
        try:
          myfile = open(file_path+filename, "rt")
          contents = myfile.read()
          fileTXT.append(contents)
        except Exception:
            print('Error reading .txt file :' + filename)

  return fileTXT
```

resumeTxt is a list containing the résumés of all the candidates.

```
# Calling the function read_files to read all the resumes
resumeTxt = read_files(resume_path)
# Displaying the first resume
resumeTxt[0]
```

Figure 5-2 shows the output.

```
'Aman Sharma \n\nCampus Address                    aman.sharma2016@vitstudent.ac.in \nMHA, VIT Che
nnai \nChennai, \nTamil Nadu  \n \n \n\nCell-7550171006 \n\nPermanent Address \n6/E, Street-27, \nSecto
r-1, Bhilai, \nChhattisgarh \n\n \n\n \n\n \nOBJECTIVE \n \nEDUCATION \n \n \n \n \n \n \n \n \n \n \n
\nDATE OF BIRTH \n \nLANGUAGES \nKNOWN \n \nCOURSEWORK \n \n \n \n \nCOMPUTER \nSKILLS \n \n, \n \nEXPE
RIENCE \n \n \n \n \n \n \n \n \n \n \n \n \nACTIVITIES \n\nSeeking to leverage my technical knowledge
to grow in your company. \n\nVellore Institute of Technology, Chennai (T.N.) \nBachelor of Computing Sc
ience and Engineering \nCGPA: 9.27/10.0 (Till III semester), to be awarded May 2020 \n \nSenior Seconda
ry School, Sector-10, Bhilai (C.G) \nHSC (Class XII), CBSE \nPercentage: 86.80, awarded May 2015 \n\nSe
nior Secondary School, Sector-10, Bhilai(C.G.) \nSSC (Class XII), CBSE \nCGPA: 10/10, awarded May 2013
\n\n10-October-1997 \n\nHindi and English \n\nObject Oriented Programming ...'
```

Figure 5-2. *Shows the output of resumeTxt*

jdTxt is a list of all the job descriptions.

```
# Calling the function read_files to read all the JDs
jdTxt = read_files(jd_path)
```

```
# Displaying the first Job Description
jdTxt[0]
```

Figure 5-3 shows the output.

```
'Position \n\nDesignation \n\nDepartment \n\nJob Location \n\nPosition Type  \n\nJob Description \n\nDe
veloper - JAVA \n\nDeveloper - JAVA \n\nMAP IT  \n\nBhopal \n\n Contractual upto 2 years. Extendable
basis performance \n\n \n\n \n  Job Objective \nObjective  of  this  job  is  to  part  of  development
teams  (open-source)  for  IT  projects;  Evaluate \nTechnical  architectures  and  provide  recommenda
tions  ;  collaborate  with  DCO  /CT  teams  at  data \ncentre,  software  development  teams,  Infrastruc
ture outsourcing team, DBAs , network administrator \nas  a  consortium       ;  Participate  &  own  da
ta  centre  upkeep  and  expansion  plans,  Liaison  with  user \ndepartments  for  state  data  centre
usage,  upkeep  of  client  assets,  own  &  drive  Archival/  back  up/ \nRestoration  policies,  Benc
hmark  new  technologies;  work  on  continuous  upgradation  of  technical \nartifacts. \n\n \n\nCore
Responsibilities: \n\n\uf0b7  Software  development  team  member  ;  ...'
```

Figure 5-3. *Shows the output of jdTxt*

Text Processing

You must process text to remove the noise and other irrelevant information. Let's follow some basic text-cleaning procedures, like lowercasing all the words, removing special characters (%, $, #, etc.), eliminating words that contain numbers (e.g., hey199, etc.), removing extra spaces, removing stop words (e.g., repetitive words like *is, the, an*, etc.), and so on.

```
# This function helps us in removing stopwords, punctuation marks, special
characters, extra spaces and numeric data. It also lowercases all the text.
Def preprocessing(Txt):
  sw = stopwords.words('english')
  space_pattern = '\s+'
  special_letters = "[^a-zA-Z#]"
  p_txt = []
  for resume in Txt:
    text = re.sub(space_pattern, ' ', resume) # Removes extra spaces
    text = re.sub(special_letters, ' ', text) # Removes special characters
    text = re.sub(r'[^\w\s]','',text) # Removes punctuations
    text = text.split() # Splits the words in a text
    text = [word for word in text if word.isalpha()] # Keeps alphabetic word
    text = [w for w in text if w not in sw] # Removes stopwords
    text = [item.lower() for item in text] # Lowercases the words
```

```
    p_txt.append(" ".join(text)) # Joins all the words   back
  return p_txt
```

"p_résuméTxt" contains all the preprocessed résumés.

```
# Calling the function preprocessing to to clean all the resumes
p_resumeTxt = preprocessing(resumeTxt)
# Displaying the first pre-processed resume
p_resumeTxt[0]
```

Figure 5-4 shows the output.

```
'aman sharma campus address aman sharma vitstudent ac mha vit chennai chennai tamil nadu cell permanent
address e street sector bhilai chhattisgarh objective education date of birth languages known coursewor
k computer skills experience activities seeking leverage technical knowledge grow company vellore insti
tute technology chennai t n bachelor computing science engineering cgpa till iii semester awarded may s
enior secondary school sector bhilai c g hsc class xii cbse percentage awarded may senior secondary sch
ool sector bhilai c g ssc class xii cbse cgpa awarded may october hindi english object oriented program
ming digital logic design database management systems data structures algorithms discrete mathematics s
oftware engineering computer architecture statistics engineer languages skills python c c r beginner ht
ml php javascript sql data structures developer vit vibrance website team participated web hackathon co
nducted mozilla research area clustering categorical data presented paper...'
```

Figure 5-4. *Shows the output of p_résuméTxt*

"jds" contains all the preprocessed job description.

```
# Calling the function preprocessing to to clean all the job description
jds = preprocessing(jdTxt)
# Displaying the first pre-processed job description
Jds[0]
```

Figure 5-5 shows the output.

'position designation department job location position type job description developer java developer ja
va map it bhopal contractual upto years extendable basis performance job objective objective job part d
evelopment teams open source it projects evaluate technical architectures provide recommendations colla
borate dco ct teams data centre software development teams infrastructure outsourcing team dbas network
administrator consortium participate data centre upkeep expansion plans liaison user departments state
data centre usage upkeep client assets drive archival back restoration policies benchmark new technolog
ies work continuous upgradation technical artifacts core responsibilities software development team mem
ber interface product principles e spoc single point contact potential candidate shall working entire s
dlc framework ide integrated development environments exposure estimation techniques diverse experience
utilizing java tools business web client server environments including j...'

Figure 5-5. *Shows the output of JDs*

You can clearly see the difference between the preprocessed and original text. Once there is processed text, it is important to convert the text into features that a machine can understand.

Text to Features

TF-IDF measures how important a word is to a particular document.

Here we are using the default TfidfVectorizer library from sklearn.

```
# Combining Resumes and Job Description for finding TF-IDF and Cosine
Similarity
TXT = p_resumeTxt+jds
# Finding TF-IDF score of all the resumes and JDs.
tv = TfidfVectorizer(max_df=0.85,min_df=10,ngram_range=(1,3))

# Converting TF-IDF to a DataFrame
tfidf_wm = tv.fit_transform(TXT)
tfidf_tokens = tv.get_feature_names()
df_tfidfvect1 = pd.DataFrame(data = tfidf_wm.toarray(),columns = tfidf_
tokens)

print("\nTD-IDF Vectorizer\n")
print(df_tfidfvect1[0:10])
```

Figure 5-6 shows the output.

TD-IDF Vectorizer

	ability	acceptance	...	xslt	years experience
0	0.000000	0.000000	...	0.000000	0.000000
1	0.033642	0.034728	...	0.000000	0.015811
2	0.030404	0.094157	...	0.000000	0.000000
3	0.000000	0.000000	...	0.017865	0.000000
4	0.000000	0.049453	...	0.000000	0.015010
5	0.000000	0.000000	...	0.037084	0.006113
6	0.010359	0.021387	...	0.000000	0.009737
7	0.060280	0.000000	...	0.000000	0.000000
8	0.006309	0.006512	...	0.000000	0.005930
9	0.027994	0.014449	...	0.000000	0.013157

[10 rows x 988 columns]

Figure 5-6. *Output of TF-IDF*

This might lead to a huge matrix based on the size of the documents, so it is important to reduce the dimensionality by using feature reduction techniques.

Feature Reduction

Truncated *singular value decomposition* (SVD) is one of the most famous approaches for dimension reduction. It factors the matrix M into the three matrixes U, Σ, and V^T to generate features. U, Σ, and V^T are described as follows.

- U is the left singular matrix
- Σ is the diagonal matrix
- V^T is the right singular matrix

Truncated SVD produces a factorization where we can specify the number of columns that we want, keeping the most relevant features.

```
# Defining transformation
dimrec = TruncatedSVD(n_components=30, n_iter=7, random_state=42)
transformed = dimrec.fit_transform(df_tfidfvect1)
```

```
# Converting transformed vector to list
vl = transformed.tolist()
# Converting list to DataFrame
fr = pd.DataFrame(vl)
print('SVD Feature Vector')
print(fr[0:10])
```

Figure 5-7 shows the output.

```
SVD Feature Vector
         0         1         2  ...        27        28        29
0  0.124064  0.068055  0.240050  ... -0.007049 -0.006502  0.003656
1  0.362082  0.553922  0.193744  ... -0.025797  0.003932  0.037391
2  0.456873  0.654845 -0.193695  ...  0.069098  0.007034  0.042495
3  0.726913 -0.319597  0.008473  ...  0.013895 -0.014235 -0.003603
4  0.435430  0.705447 -0.178258  ...  0.084570  0.149095 -0.013346
5  0.832300 -0.296318 -0.006792  ... -0.112076  0.227854  0.165388
6  0.494776  0.569386 -0.105892  ...  0.080577  0.009561 -0.060952
7  0.258640  0.376127  0.554150  ... -0.015597  0.001795 -0.003945
8  0.785569 -0.363791 -0.019028  ... -0.096952 -0.019138 -0.102783
9  0.472396  0.683913 -0.172216  ...  0.041386  0.049625  0.022959

[10 rows x 30 columns]
```

Figure 5-7. *Output of SVD*

Model Building

A similarity measure problem in which the top N (N is customizable) matching résumés are recommended to the recruiter. For this, let's use cosine similarity, which measures how similar the two vectors are.

The formula is as follows.

$$cos(x, y) = \frac{x \cdot y}{\|x\| * \|y\|}$$

- $x \cdot y$ is the dot product of vectors x and y

- $||x||$ and $||y||$ is the length of two vectors x and y

- $||x|| * ||y||$ is the cross product of two vectors x and y

```
# Calculating Cosine Similarity between JDs and resumes to find out which
resume is the best fit for a job description
similarity = cosine_similarity(df_tfidfvect1[0:len(resumeTxt)],df_
tfidfvect1[len(resumeTxt):])

# Column names for job description
abc = []
for i in range(1,len(jds)+1):
  abc.append(f"JD {i}")
# DataFrame of similarity score
Data=pd.DataFrame(similarity,columns=abc)
print('\nCosine Similarity\n')
print(Data[0:10])
```

Figure 5-8 shows the output.

```
Cosine Similarity
```

	Java Developer	Project Manager	Business Analyst
0	0.107977	0.180878	0.068128
1	0.164169	0.264543	0.265139
2	0.126303	0.182786	0.341395
3	0.408364	0.047575	0.053543
4	0.123725	0.224996	0.371208
5	0.428426	0.080865	0.084784
6	0.235509	0.239328	0.285766
7	0.106679	0.387226	0.167779
8	0.404402	0.050072	0.073850
9	0.126238	0.222074	0.433800

Figure 5-8. *The output displays the similarity score between each résumé(representing each row) and respective job description*

Extracting Entities

Now that we have résumé similarity scores with respect to each of the job descriptions, let's extract required entities like candidate name, phone number, e-mail address, skills, year of experience, and former employers.

Let's formulate regex to extract phone numbers:

```python
# DataFrame of original resume
t = pd.DataFrame({'Original Resume':resumeTxt})
dt = pd.concat([Data,t],axis=1)

# Function to find phone numbers
def number(text):

        # compile helps us to define a pattern for matching it in the text
        pattern = re.compile(r'([+(]?\d+[)\-]?[ \t\r\f\v]*[(]?\d{2,}[()\-
        ]?[ \t\r\f\v]*\d{2,}[()\-]?[ \t\r\f\v]*\d*[ \t\r\f\v]*\d*[ \t\r\
        f\v]*)')

        #  findall finds the pattern defined in compile
        pt = pattern.findall(text)

        #  sub replaces a pattern matching in the text
        pt = [re.sub(r'[,.]', '', ah) for ah in pt if len(re.sub(r'[()\-.,\
        s+]', '', ah))>9]
        pt = [re.sub(r'\D$', '', ah).strip() for ah in pt]
        pt = [ah for ah in pt if len(re.sub(r'\D','',ah)) <= 15]

        for ah in list(pt):
            #  split splits a text
            if len(ah.split('-')) > 3: continue

            for x in ah.split("-"):

                try:

                    #  isdigit checks whether the text is number or not
                    if x.strip()[-4:].isdigit():
                        if int(x.strip()[-4:]) in range(1900, 2100):

                            #  removes a the mentioned text
```

```
                    pt.remove(ah)
            except: pass

    number = None
    number = list(set(pt))
    return number
```

dt['Phone No.'] contains all the phone numbers of a candidate.

```
# Calling the number function to get the list of candidate's numbers
dt['Phone No.']=dt['Original Resume'].apply(lambda x: number(x))
print("Extracting numbers from dataframe columns:")
dt['Phone No.'][0:5]
```

Figure 5-9 shows the output.

```
Extracting numbers from dataframe columns:
0           [7550171006]
1         [281-212-3592]
2         [970-294-9622]
3       [(669) 261-3506]
4         [703-743-0795]
Name: Phone No., dtype: object
```

Figure 5-9. *Shows first five numbers extracted from résumé*

Now let's extract each candidate's e-mail address using regex.

```
# Here we are extracting e-mail from the resumes
def email_ID(text):

    # compile helps us to define a pattern for matching it in the text
    r = re.compile(r'[A-Za-z0-9_.+-]+@[a-zA-Z0-9-]+\.[a-zA-Z0-9-.]+')

    return r.findall(str(text))
```

dt['E-Mail ID'] contains the candidates' e-mail addresses.

```
#  Calling the email_ID function to get the list of candidate's e-mails
dt['E-Mail ID']=dt['Original Resume'].apply(lambda x: email_ID(x))
print("Extracting e-mail from dataframe columns:")
dt['E-Mail ID'] [0:5]
```

Figure 5-10 shows the output.

```
Extracting e-mail from dataframe columns:
0                    [aman.sharma2016@vitstudent.ac.in]
1                             [adhigopalam@gmail.com]
2                    [danny@stackitprofessionals.com]
3                          [bsudabathula@gmail.com]
4      [amirindersingh1234@gmail.com, Praveen@indique...
Name: E-Mail ID, dtype: object
```

Figure 5-10. *Shows first five e-mail extracted from résumé*

Next, we remove the phone number and e-mail addresses of the candidates from résumé corpus because we want to extract the years of experience and name of the candidates, and the integers in phone numbers can be misinterpreted as years of experience or the function for recognizing the candidates' names.

```
# Function to remove phone numbers to extract the year of experience and
name of a candidate
def rm_number(text):
    try:

        # compile helps us to define a pattern for matching it in the text
        pattern = re.compile(r'([+(]?\d+[)\-]?[ \t\r\f\v]*[(]?\d{2,}[()\-
        ]?[ \t\r\f\v]*\d{2,}[()\-]?[ \t\r\f\v]*\d*[ \t\r\f\v]*\d*[ \t\r\
        f\v]*)')

        #  findall finds the pattern defined in compile
        pt = pattern.findall(text)

        #  sub replaces a pattern matching in the text
        pt = [re.sub(r'[,.]', '', ah) for ah in pt if len(re.sub(r'[()\-.,\
        s+]', '', ah))>9]
        pt = [re.sub(r'\D$', '', ah).strip() for ah in pt]
        pt = [ah for ah in pt if len(re.sub(r'\D','',ah)) <= 15]

        for ah in list(pt):
            #  split splits a text
```

```
           if len(ah.split('-')) > 3: continue

           for x in ah.split("-"):

               try:

                   # isdigit checks whether the text is number or not
                   if x.strip()[-4:].isdigit():
                       if int(x.strip()[-4:]) in range(1900, 2100):
                   # removes a the mentioned text
                           pt.remove(ah)
               except: pass
       number = None
       number = pt
       number = set(number)
       number = list(number)
       for i in number:
         text = text.replace(i," ")
       return text
   except:
       pass
```

dt['Original'] contains all the résumés after removing phone number of a candidate.

```
# Calling the function rm_number to remove the phone number
dt['Original']=dt['Original Resume'].apply(lambda x: rm_number(x))
```

Now let's remove e-mail addresses.

```
# Function to remove emails to extract the year of experience and name of a
candidate
def rm_email(text):

   try:
       email = None

       # compile helps us to define a pattern for matching it in the text
       pattern = re.compile('[\w\.-]+@[\w\.-]+')

       #  findall finds the pattern defined in compile
```

```
    pt = pattern.findall(text)

    email = pt
    email = set(email)
    email = list(email)

    for i in email:
      # replace will replace a given string with another
      text = text.replace(i," ")
    return text

  except:
    pass
```

dt['Original'] contains all the résumés after removing a candidate's phone number and e-mail address.

```
# Calling the function rm_email to remove the e-mails
dt['Original']=dt['Original'].apply(lambda x: rm_email(x))
print("Extracting numbers from dataframe columns:")
dt['Original'][0:5]
```

Figure 5-11 shows the output.

```
Resumes with emails and numbers removed :
0     Aman Sharma \n\nCampus Address            ...
1     Adhi Gopalam\n\n \n\n \n\n\n\nCertified Scrum ...
2     Hyma Gumpu\n\n     \n\nContact No:  \n\nProfile...
3      Name: Balakrishna Sudabathula\n\n  Email:  \n...
4     Amrinder Pelia                            ...
Name: Original, dtype: object
```

Figure 5-11. *Shows first five résumés with numbers and e-mail removed*

Now that we have removed e-mail addresses and phone numbers, let's extract the candidate names using POS tagging. For more information on POS tagging in NLP, refer to our book *Natural Language Processing Recipes: Unlocking Text Data with Machine Learning and Deep Learning Using Python* (Apress, 2019).

```python
# Function to extract candidate name
def person_name(text):
    # Tokenizes whole text to sentences
    Sentences = nltk.sent_tokenize(text)
    t = []
    for s in Sentences:
        # Tokenizes sentences to words
        t.append(nltk.word_tokenize(s))
    # Tags a word with its part of speech
    words = [nltk.pos_tag(token) for token in t]
    n = []
    for x in words:
        for l in x:
            # match matches the pos tag of a word to a given tag here
            if re.match('[NN.*]', l[1]):
                n.append(l[0])
    cands = []
    for nouns in n:
        if not wordnet.synsets(nouns):
                cands.append(nouns)
    cand = ' '.join(cands[:1])
    return cand
```

dt['Candidate\'s Name'] contains names of all the candidates.

```python
# Calling the function name to extract the name of a candidate
dt['Candidate\'s Name']=dt['Original'].apply(lambda x: person_name(x))
print("Extracting names from dataframe columns:")
dt['Candidate\'s Name'][0:5]
```

Figure 5-12 shows the output.

143

```
Extracting names from dataframe columns:
0              Aman
1              Adhi
2              Hyma
3       Balakrishna
4          Amrinder
Name: Candidate's Name, dtype: object
```

Figure 5-12. *Shows first five candidate names extracted from résumé*

Now, let's extract years of experience using regex again.

```
#function to find the years of experience
def exp(text):
    try:
        e = []
        p = 0
        text  = text.lower()
        # Searches a pattern text string similar to the given pattern
        pt1 = re.search(r"(?:[a-zA-Z'-]+[^a-zA-Z'-]+){0,7}experience(?:
        [^a-zA-Z'-]+[a-zA-Z'-]+){0,7}", text)
        if(pt1 != None):
            # groups all the string found in match
            p = pt1.group()
        # Searches a pattern text string similar to the given pattern
        pt2 = re.search(r"(?:[a-zA-Z'-]+[^a-zA-Z'-]+){0,2}year(?:[^a-zA-Z'-
        ]+[a-zA-Z'-]+){0,2}", text)
        if(pt2 != None):
            # groups all the string found in match
            p = pt2.group()
        # Searches a pattern text string similar to the given pattern
        pt3 = re.search(r"(?:[a-zA-Z'-]+[^a-zA-Z'-]+){0,2}years(?:[^a-zA-
        Z'-]+[a-zA-Z'-]+){0,2}", text)
        if(pt3 != None):
            # groups all the string found in match
            p = pt3.group()
        if(p == 0):
```

```
        return 0
    #  findall finds the pattern defined in compile
    ep = re.findall('[0-9]{1,2}',p)
    ep_int = list(map(int, ep))
    # this for loop is for filtering and then appending string
    containing years of experience
    for a in ep:
        for b in ep_int:
            if len(a) <= 2 and b < 30:
                e.append(a)
    ep = ''.join(e[0])
    #  findall finds the pattern defined in compile
    p1 = re.findall('[0-9]{1,2}.[0-9]{1,2}',p)
    exp = []
    if not p1:
        exp.append(ep)
        exp = ''.join(ep)
    else:
        exp.append(p1)
        exp = ''.join(p1)
except:
    exp=0
return exp
```

dt['Experience'] contains years of experience of all the candidates.

```
#Calling the function exp to extract the year of experience of the
candidate
dt['Experience']=dt['Original'].apply(lambda x: exp(x))
print("Extracting e-mail from dataframe columns:")
dt['Experience']
```

Figure 5-13 shows the output.

```
Extracting years of experience from dataframe column:
0     0
1     12
2     7
3     10
4     10
Name: Experience, dtype: object
```

Figure 5-13. *Shows first five candidate years of experience extracted from résumé*

Extract skills using a predefined skill set.

```
# Importing a file of pre-defined skills & Converting DataFrame to list
skills = pd.read_excel('/content/drive/MyDrive/skills.xlsx')
skills = skills.values.flatten().tolist()
i = 0
skill = []
for z in skills:
    r = z.lower()
    skill.append(r)
    i += 1
```

Map the skills from the résumé to a predefined skill set and extract them.

```
# Function to extract skills from candidate's resume
def skills(text):
    sw = set(nltk.corpus.stopwords.words('english'))
    tokens = nltk.tokenize.word_tokenize(text)
    # remove the punctuation
    ft = [w for w in tokens if w.isalpha()]
    # remove the stop words
    ft = [w for w in tokens if w not in sw]
    # generate bigrams and trigrams (like Machine Learning)
    n_grams = list(map(' '.join, nltk.everygrams(ft, 2, 3)))
    fs = set()
    # we text for each token in our skills database
    for token in ft:
        if token.lower() in skill:
```

```
        fs.add(token)
    # we text for each bigram and trigram in our skills database
    for ngram in n_grams:
        if ngram.lower() in skill:
            fs.add(ngram)
    return fs
```

dt['Skills'] contains skills of all the candidates.

```
# Calling the function skills to extract the skills of a candidate
dt['Skills']=dt['Original'].apply(lambda x: skills(x))
print("Extracting Person Name from dataframe columns:")
dt['Skills']
```

Figure 5-14 shows the output.

```
    Extracting skills from dataframe columns:
    0     {Database Management, Data Structures, SQL, CO...
    1     {project management, Java, Word, Excel, Projec...
    2     {project management, Business Analysis, SQL, p...
    3           {Angular, SQl, SQL, Java, Java Script, JAVA}
    4     {GAP Analysis, project management, SQL, Java, ...
    Name: Skills, dtype: object
```

Figure 5-14. *Shows first five candidate skills extracted from résumé*

Location is another important parameter for a recruiter to understand which location the candidate is from, before proceeding with the candidature. Let's extract location using entity extractor.

```
# Function to extract Location
def location(text):

  place_entity = locationtagger.find_locations(text=text)
  return place_entity.cities
```

dt['Location'] contains location of all the candidates.

```
# Calling the function location to extract the location of a candidate
dt['Location']=dt['Resume'].apply(lambda x: location(x))
```

```
print("Extracting cities from dataframe columns:")
dt['Location']
```

Figure 5-15 shows the output.

```
Extracting location from dataframe columns:
0                                    [Chennai, Bhilai]
1       [Richmond, Naperville, Salem, Pittsburgh, Balt...
2       [New York, Los Angeles, Rendezvous, Case, Rad,...
3       [Mumbai, Hyderabad, Sunnyvale, Spring, Singlet...
4       [Gap, As, Jackson, Norristown, Cincinnati, Ban...
Name: Location, dtype: object
```

Figure 5-15. *Shows first five candidate locations extracted from résumé*

Similarly, let's extract the company name using *named-entity recognition* (NER). For more information on NER, refer to our book *Natural Language Processing Recipes: Unlocking Text Data with Machine Learning and Deep Learning Using Python* (Apress, 2019).

```
#Function to extract Company name
def CompanyName(text):
    #  for tagging each entity with its it's labels
    tokens = nlp(str(text))

    x=[]
    #  for loop for extracting company names
    for ent in tokens.ents:

      if ent.label_ == 'ORG':
        return ent.text
```

dt['Company Name'] contains location of all the candidates.

```
# Calling the function CompanyName to extract past companies of a candidate
dt['Company Name']=dt['Original Resume'].apply(lambda x: CompanyName(x))

print("Extracting Person Name from dataframe columns:")
dt['Company Name']
```

Figure 5-16 shows the output.

```
Extracting Company Name from dataframe columns:
0        Vellore Institute of Technology
1                            Infosys Ltd.
2                         Volvo Group Pvt
3                         HCL Technologies
4        Tata Consultancy Services Pvt
Name: Company Name, dtype: object
```

Figure 5-16. *Shows first five candidates past company extracted from résumé*

Ranking

Now that we have extracted all the important information regarding the candidates, we display a rank list of candidate résumés for all different job descriptions. The items can now be ordered by simply arranging them in descending order with respect to the similarity score of a given job description.

The following is the final result for the project manager profile.

```
# Final result for Project Manager profile
pm = dt[['Project Manager','Candidate\'s Name','Phone No.','E-Mail
ID','Skills','Experience', 'Location','Company Name']]
pm=pm.sort_values(by='Project Manager',ascending=False)
pm[0:10]
```

Figure 5-17 shows the output.

	Project Manager	Candidate's Name	Phone No.	E-Mail ID	Skills	Experience	Location	Company Name
17	0.448199	Priya	[+1 503-536-2757]	[manjunath@nichesoftsolutions.com]	[Java Servlets, Microsoft Access, Risk Assessm...	7	[Mumbai, Delhi, Gurgeon, London, Vas, Mena, Me...	Daimler Company
7	0.387226	Jagan	[+1 (248) 412-1658. (248) 412-1658]	[jagansrconsult@gmail.com, jagansrconsult@gmai...	[Information Technology, Cloud Computing, Big ...	20	[Chennai, Change, Riyadh, Chicago, New York, P...	Tata Motors LTD
19	0.295847	Ajay	[+1 (847)899-4194]	[ajaydt@gmail.com]	[Project management, risk assessment, GAP Anal...	14	[Kolkata, Sale, New York, Pittsburgh, Singapor...	LIC Company
18	0.284137	Mahesh	[+18327758735]	[asmith@nichesoftsolutions.com]	[organizational skills, Project Planning, soft...	17	[Bangalore, March, Lenexa, Core, Usa, Leader,...	TCS
1	0.264543	Adhi	[281-212-3592]	[adhigopalam@gmail.com]	[project management, Java, Word, Excel, Projec...	12	[Richmond, Naperville, Salem, Pittsburgh, Balt...	Infosys Ltd.
27	0.260352	Shail	[646442 9382]	[shailjtank@gmail.com]	[Project Management, database design, Decision...	9	[Gap, London, March, As, Minneapolis, Parsippa...	Sony Company
6	0.239328	Kashyep	[201-532-6397]	[kashyapkvora@gmail.com]	[SQL, Business Analysis, gap analysis, data mi...	10	[Mumbai, Gap, Warren, Baltimore, Richardson, U...	SAP
4	0.224996	Amrinder	[703-743-0795]	[amrindersingh1234@gmail.com, Praveen@indique...	[GAP Analysis, project management, SQL, Java,...	10	[Gap, As, Jackson, Norristown, Cincinnati, Ban...	Tata Consultancy Services Pvt
9	0.222074	Mounika	[414-909-0756]	[Mounika10200@gmail.com]	[Operating Systems, analytical skills, Informa...	8	[Gap, Moscow, As, Lead, Strong, China, Us, Atl...	China Life Insurance company and Information
24	0.211687	Madhuri	[484 754 6872]	[madurichand@gmail.com]	[risk assessment, GAP Analysis, database desig...	7	[Hyderabad, Wilmington, Anchorage, Bloomfield,...	AmerisourceBergen Company

Figure 5-17. *Shows tabular representation of top 10 résumés for the project manager position*

The Project Manager column indicates the similarity score of a particular résumé with the job description for a project manager.

The following is the final result for the business analyst profile.

```
# Final Result for Business Analyst
ba = dt[['Business Analyst','Candidate\'s Name','Phone No.','E-Mail
ID','Skills','Experience', 'Location','Company Name']]
ba=ba.sort_values(by='Business Analyst',ascending=False)
ba[0:10]
```

Figure 5-18 shows the output.

	Business Analyst	Candidate's Name	Phone No.	E-Mail ID	Skills	Experience	Location	Company Name
9	0.433800	Mounika	[414-909-0756]	[Mounika10200@gmail.com]	[Operating Systems, analytical skills, Informa...	8	[Gap, Moscow, As, Lead, Strong, China, Us, Atl...	China Life Insurance company and Information
24	0.391456	Madhuri	[464 754 6872]	[madurichand@gmail.com]	[risk assessment, GAP Analysis, database desig...	7	[Hyderabad, Wilmington, Anchorage, Bloomfield,...	AmerisourceBergen Company
15	0.385279	Dhanalaxmi	[916-282-9259]	[sam@vishconsultingservices.com]	[Project Planning, Business Analysis, SQL, gap...	7	[Pune, Change, Tampa, Jacksonville, Sacramento...	IBM Company
4	0.371208	Amrinder	[703-743-0795]	[amrindersingh1234@gmail.com, Praveen@indique...	[GAP Analysis, project management, SQL, Java,...	10	[Gap, As, Jackson, Norristown, Cincinnati, Ban...	Tata Consultancy Services Pvt
2	0.341395	Hyma	[970-294-9622]	[danny@stackitprofessionals.com]	[project management, Business Analysis, SQL, p...	7	[New York, Los Angeles, Rendezvous, Case, Rad,...	Volvo Group Pvt
27	0.324604	Shail	[646442 9382]	[shailjtank@gmail.com]	[Project Management, database design, Decision...	9	[Gap, London, March, As, Minneapolis, Parsippa...	Sony Company
19	0.324550	Ajay	[+1 (847)899-4194]	[ajaydt@gmail.com]	[Project management, risk assessment, GAP Anal...	14	[Kolkata, Sale, New York, Pittsburgh, Singapor...	LIC Company
6	0.285766	Kashyap	[201-532-6397]	[kashyapkvora@gmail.com]	[SQL, Business Analysis, gap analysis, data mi...	10	[Mumbai, Gap, Warren, Baltimore, Richardson, U...	SAP
17	0.285576	Priya	[+1 503-536-2757]	[manjunath@nichesoftsolutions.com]	[Java Servlets, Microsoft Access, Risk Assessm...	7	[Mumbai, Delhi, Gurgaon, London, Vas, Mena, Me...	Daimler Company
13	0.275595	A N DINESHKUMARR	[9944819797, +91 9538820896]	[andineshkumarr@yahoo.com]	[Business Analysis, SQL]	3	[Bangalore, Chennai, Hyderabad, London, Geneva...	Morgan stanley company

Figure 5-18. *Shows tabular representation of top 10 résumés for the business analyst position*

The Business Analyst column indicates the similarity score of a particular résumé with a job description for a business analyst.

The following is the final result for the Java developer profile.

```
# Final Result for Java Developer
jad = dt[['Java Developer','Candidate\'s Name','Phone No.','E-Mail
ID','Skills','Experience', 'Location','Company Name']]
jad=jad.sort_values(by='Java Developer',ascending=False)
jad[0:10]
#Here "Java Developer" column indicates similarity score of that particular
resume with Java Developer JD
```

Figure 5-19 shows the output.

	Java Developer	Candidate's Name	Phone No.	E-Mail ID	Skills	Experience	Location	Company Name
20	0.459094	Siddharth	[(646) 494-6583]	[siddhartha.g225@gmail.com]	(Angular, Data structures, Big Data, debugging…	9	[Bangalore, As, San Francisco, Spring, Fort Wo…	Volvo Company
25	0.451500	Manohar	[(614) 285-5486]	[manohar.r7754@gmail.com]	(Angular, java, Operating Systems, cloud compu…	4	[As, Spring, Atlanta, East Hanover, Newark, Mo…	Samsung Company
12	0.450859	VARUN	[601 651 0507]	[Varunkumar.work@gmail.com601]	(Angular, java, Cloud Computing, project manag…	8	[Spring, Dallas, Enterprise, Eureka, Page, Kum…	SAP Company
26	0.438464	KIRAN	[510-770-6277]	[kumarjava174@gmail.com]	(Angular, java, contract negotiation, Operatin…	8	[Malvern, Jar, Chicago, Houston, Spring, Mobil…	Accenture Company
10	0.437803	Nithin	[214-509-7784]	[nithin7445@gmail.com]	(java, Angular, Java Servlets, Information Tec…	8	[Jar, Spring, Austin, San Antonio, Columbia, P…	Amazon Company
11	0.433409	Kumar	[469 524 9207]	[naveenkumar.java18@gmail.com]	(java, Angular, Operating Systems, Data struct…	8	[Bangalore, Spring, Dallas, Enterprise, Eureka…	Google Company
14	0.429513	Avinash	[913-730-0694]	[Avinash87.java@gmail.com]	(Angular, java, Java Servlets, debugging, SQL,…	8	[Hyderabad, New York, Sunnyvale, Spring, Trent…	SAP Company
5	0.428426	AMULYA	[(515)309-1612]	[amulya.javadeveloper@gmail.com]	(java, Angular, Customer Service, project mana…	8	[Hyderabad, Jar, Chicago, New York, Spring, Ka…	Goldman Sachs
21	0.425585	Sudhakar	[201-535-3300]	[psrleo81@gmail.com]	(java, Angular, organizational skills, Data St…	9	[Washington, New York, Naperville, Spring, Sta…	SAP Company
3	0.408364	Balakrishna	[(669) 261-3506]	[bsudabathula@gmail.com]	(Angular, SQI, SQL, Java, Java Script, JAVA)	10	[Mumbai, Hyderabad, Sunnyvale, Spring, Singlet…	HCL Technologies

Figure 5-19. *Shows tabular representation of Top 10 résumés for the position of Java Developer*

Visualization

Now that we have found the top 10 résumés, let's make a word cloud of the words present in the top matching résumé for the given job description. This helps the recruiter to further validate candidates by glimpsing a résumé based on the job description.

```
# Create and generate a word cloud image for the best candidate for Project
Manager
wordcloud = WordCloud(width = 800, height = 500,background_color ='white'
,min_font_size = 10).generate(resumeTxt[17])

# Display the generated image
plt.figure(figsize = (20, 5), facecolor = None)
plt.imshow(wordcloud)
```

```
plt.axis("off")
plt.tight_layout(pad = 0)
plt.show()
```

Figure 5-20 shows the output.

Figure 5-20. *This is a WordCloud for résumé 17 index as it is the top résumé for project manager profile*

The résumé contains words like *management, plan, scope, process, delivery, requirements,* and *risk,* which are relevant for the project manager profile. This shows that our model is doing a good job.

```
# Create and generate a word cloud image for the best candidate for
Business Analyst
wordcloud = WordCloud(width = 800, height = 500,background_color ='white',
min_font_size = 10).generate(resumeTxt[12])

# Display the generated image
plt.figure(figsize = (20, 5), facecolor = None)
plt.imshow(wordcloud)
plt.axis("off")
plt.tight_layout(pad = 0)
plt.show()
```

Figure 5-21 shows the output.

Figure 5-21. *A word cloud for résumé 12 index. It is the top résumé for business analyst profile*

```
# Create and generate a word cloud image for the best candidate for
Java Developer
wordcloud = WordCloud(width = 800, height = 500,background_color ='white',
min_font_size = 10).generate(resumeTxt[23])
```

```
# Display the generated image
plt.figure(figsize = (20, 5), facecolor = None)
plt.imshow(wordcloud)
plt.axis("off")
plt.tight_layout(pad = 0)
plt.show()
```

Figure 5-22 shows the output.

Figure 5-22. *A word cloud for résumé 23 index—the top résumé for Java developer profile*

Similarly, important words from this résumé are Java, CSS, app, framework, and so on which are relevant given the Java developer profile.

Let's visualize and compare using a word cloud of the best candidate's résumé. The model identifies a Java developer by job description.

```
# Create a world cloud of job description of a Java Developer
wordcloud = WordCloud(width = 800, height = 500,background_color
='white',min_font_size = 10).generate(jds[0])

# Display the generated image
plt.figure(figsize = (20, 5), facecolor = None)
plt.imshow(wordcloud)
plt.axis("off")
plt.tight_layout(pad = 0)
plt.show()
```

Figure 5-23 shows the output.

Figure 5-23. *A word cloud for the job description of Java developer*

Wow, there are a lot of similar skills present in both the word clouds. Similarly, you can compare with the rest of the job descriptions and respective top résumés.

Conclusion

We implemented a basic version of the AI-based résumé screening and shortlisting model and got a sensible output. Please note that there is a lot you can do on top of this to scale it further. The focus is not to get the great output but to understand how to approach the problem and the steps we need to follow. Let's move on to more exciting projects in the upcoming chapters.

## CHAPTER 6

Creating an E-commerce Product Categorization Model Using Deep Learning

This chapter explores multiclass classification using deep learning. You look at different deep neural networks, like CNN, RNN, and LSTM, and ways to tune and evaluate them.

Problem Statement

In the era of e-commerce and retail, the biggest challenge is to tag a particular product or SKU to its category. There are billions of items, and manually tagging is an inefficient task and cost to a company. It must be done intelligently and fast. New items are added to the website or store every day, and detecting the category is critical. Machine learning and natural language processing come to the rescue to solve this problem and save a lot of time and money.

Figure 6-1 is an example of a category hierarchy. There are categories, subcategories, and within that, products/items. The whole structure is hierarchical.

© Akshay Kulkarni, Adarsha Shivananda and Anoosh Kulkarni 2022
A. Kulkarni et al., *Natural Language Processing Projects*, https://doi.org/10.1007/978-1-4842-7386-9_6

Tv, Appliances, Electronics	Men's Fashion
Televisions	T-Shirts
Camera	Jeans
Cell Phones	
Speakers	**Home, Kitchens**
Headphones	Furniture
Refrigerator	Dinning set

Figure 6-1. *Categories details*

In this project, we build a predictive model and categorize products in an e-commerce data set. Product categorization is a supervised classification problem where the product categories are the target classes, and the features are the words extracted from the product description or an image.

The aim is to successfully classify product categories with high precision using state machine learning and deep learning techniques.

Methodology and Approach

Product classification is one of the classic problem statements in the e-commerce industry, and there are multiple approaches to solve it. Let's use the following approaches.

- Image classification

- Text classification

The problem can be solved in two ways. There are images for every item mapped to a particular category. We need to train deep learning algorithms and use that machine to predict the category whenever there is a new image of the product. But the challenge is, not every time seller provides the image which becomes a problem.

The other approach is multiclass text classification. Instead of images, we need to consider the product description as features and product category as a label to build the classifier. The idea is that a product description has detailed information about a particular product, and with some confidence, the text present in the description can predict the category.

Figure 6-2 shows the methodology adopted for an e-commerce product data set classification using the text classification approach.

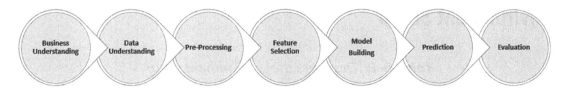

Figure 6-2. *Flow diagram*

This model can be built using traditional machine learning algorithms (naive Bayes classifier, logistics regression, support vector machine classifier, random forest classifier) and deep neural networks.

Figure 6-3 shows the types of algorithms.

Features	Naive bayes [19]	Max Entropy [19]	Boosted trees[19]	SVM[6]	Random forest[19]	KNN[15]
Based on	Bayes theorem	Feature based classifier	Decision tree Learning	Distance vector	Decision tree Aggregation	Nearest neighbor
Simplicity	Very Simple	Hard	Moderate	Moderate	Simple	Simple
Performance	Better	Good	Good	Better	Excellent	Poor
Accuracy	Good	High	Poor	Good	Excellent	Good
Memory requirement	Low	High	Low	High	High	Low
Time Required	Low	Moderate	High	Moderate	High	Very low

Figure 6-3. *Types of algorithms*

We implement the multiclass text classification using different deep neural networks algorithms, based on Keras sequential libraries, underlying back end as tensor flow framework.

Environment Setup

Table 6-1. *Describes the environment setup that was used in this project*

Setup	Version
Anaconda Distribution	5.2.0
Python	3.6.5
Notebook	5.5.0
NumPy	1.14.3
pandas	0.23.0
scikit-learn	0.19.1
Matplotlib	2.2.2
Seaborn	0.8.1
NLTK	3.3.0
Tensorflow	1.12.0
Tensorboard	1.12.0
Keras Preprocessing	1.0.5

The e-commerce product categorization data set we used has 17,533 observations and 15 attributes listed in Table 6-2. This is a multiclass classification problem with the target labeled as **product_category** _tree and description as the independent variable (Features are extracted from text). This is a free source data set and can be downloaded from https://data.world/promptcloud/product-details-on-flipkart-com.

Table 6-2. *Data details*

Attribute Name	Data Type
uniq_id	Object
crawl_timestamp	Object
product_url	object
product_name	object
Pid	object
retail_price	float64
discounted_price	float65
image	object
is_FK_Advantage_product	bool
description	object
product_rating	object
overall_rating	object
brand	object
product_specifications	object
product_category_tree	object

Understanding the Data

Before performing any processing on the available data, exploratory data analysis is recommended. This process includes visualization of the data for better understanding, identifying the outliers, and skewed predictors. These tasks help you understand and inspect the data and identify the missing values and irrelevant information in the data set.

Let's import the required libraries.

```
# Data Manipulation
import numpy as np
import pandas as pd

# Visualization
import matplotlib.pyplot as plt
```

```
import seaborn as sns

import keras
from keras.preprocessing.text import Tokenizer
from keras.models import Sequential
from keras.layers import Dense
from keras.preprocessing.sequence import pad_sequences
from keras.layers import Input, Dense, Dropout, Embedding, LSTM, Flatten,
Conv1D, MaxPooling1D
from keras.models import Model
from tensorflow.keras.utils import to_categorical
from keras.callbacks import ModelCheckpoint
from keras import layers

#NLP for text pre-processing
import nltk
from nltk.corpus import stopwords
from wordcloud import WordCloud, STOPWORDS

#·for spliting data set and metrics
from sklearn.model_selection import train_test_split
from sklearn.metrics import accuracy_score

#Handling imbalance data
from imblearn.over_sampling import SMOTE

# Plot the Figures Inline
%matplotlib inline
```

Let's import the data and understand the columns.

```
# Loading the dataset
Prod_cat_data = pd.read_csv('ecommerce.csv')

Prod_cat_data.shape
```

The following is the output.

```
(17533, 15)
```

The data set has 15 columns and 17533 observations. In reality, we might have more products than this.

```
Prod_cat_data.info()
```

The following is the output.

```
<class 'pandas.core.frame.DataFrame'>
RangeIndex: 17533 entries, 0 to 17532
Data columns (total 15 columns):
uniq_id                  17533 non-null object
crawl_timestamp          17533 non-null object
product_url              17533 non-null object
product_name             17533 non-null object
product_category_tree    17533 non-null object
pid                      17533 non-null object
retail_price             17472 non-null float64
discounted_price         17472 non-null float64
image                    17530 non-null object
is_FK_Advantage_product  17533 non-null bool
description              17532 non-null object
product_rating           17533 non-null object
overall_rating           17533 non-null object
brand                    12301 non-null object
product_specifications   17522 non-null object
dtypes: bool(1), float64(2), object(12)
```

The e-commerce data set has 15 attributes; out of these columns, we only extracted the following elements for further analysis: description and product_category_tree.

The remaining columns are not considered for building the text classification model. So, only the description column is considered useful.

Exploratory Data Analysis

Before moving ahead, let's look at the distribution of each category.

```
Prod_cat_data['product_category_tree'].value_counts()
```

The following is the output.

```
Clothing                        6198
Jewelry                         3531
Footwear                        1227
Mobiles & Accessories           1099
Automotive                      1012
Home Decor & Festive Needs       929
Kitchen & Dining                 647
Computers                        578
Watches                          530
Baby Care                        483
Tools & Hardware                 391
Toys & School Supplies           330
Pens & Stationery                313
Bags, Wallets & Belts            265
Name: product_category_tree, dtype: int64
```

Let's plot product category distribution to visualize and understand it better.

```
fig, ax = plt.subplots(figsize=[8,4], nrows=1, ncols=1)
Prod_cat_data['product_category_tree'].value_counts().plot(ax=ax,
kind='bar', title='Product Category Distribution')
```

Figure 6-4 shows the product category distribution.

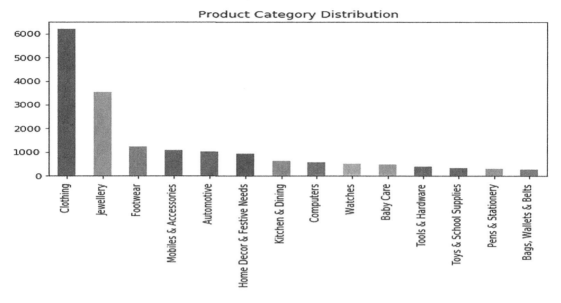

Figure 6-4. *Output*

Most products belong to Clothing or Jewelry, followed by Footwear and Mobiles & Accessories. There are very few products under Bags, Wallets & Belts.

Data Preprocessing

In this project, we are bound to do certain data preprocessing steps, including data cleaning, preparation, transformation, and dimensionality reduction.

First, we investigate the usual data set level cleaning and then later jump into text preprocessing.

Let's see if there are any missing values in the description column.

```
# Number of missing values in each column
missing = pd.DataFrame(Prod_cat_data.isnull().sum()).rename(columns = {0:
'missing'})
```

```
# Create a percentage of missing values
missing['percent'] = missing['missing'] / len(Prod_cat_data)
```

```
# sorting the values in descending order to see highest count on the top
missing.sort_values('percent', ascending = False)
```

Figure 6-5 is the output for the code.

	missing	percent
brand	5232	0.298409
retail_price	61	0.003479
discounted_price	61	0.003479
product_specifications	11	0.000627
image	3	0.000171
description	1	0.000057
length	1	0.000057
uniq_id	0	0.000000
crawl_timestamp	0	0.000000
product_url	0	0.000000
product_name	0	0.000000
product_category_tree	0	0.000000
pid	0	0.000000
is_FK_Advantage_product	0	0.000000
product_rating	0	0.000000
overall_rating	0	0.000000

Figure 6-5. *Output*

Description feature has one missing value. Let's drop that missing value observation from the data set.

```
# Removing missing values in description
Prod_cat_data=Prod_cat_data[pd.notnull(Prod_cat_data['description'])]
```

Let's look at the word distribution to understand what kind of words are present in the corpus.

```
#Adding New column with no of words in the description before text pre
processing
```

166

```
Prod_cat_data['no_of_words'] = Prod_cat_data.description.apply(lambda a :
len(a.split()))bins=[0,50,75, np.inf]
Prod_cat_data['bins']=pd.cut(Prod_cat_data.no_of_words,
bins=[0,100,300,500,800, np.inf], labels=['0-100', '100-200', '200-
500','500-800' ,'>800'])

words_distribution = Prod_cat_data.groupby('bins').size().reset_index().
rename(columns={0:'word_counts'})

sns.barplot(x='bins', y='word_counts', data=words_distribution).set_
title("Word distribution per bin")
```

Figure 6-6 shows the distribution of words.

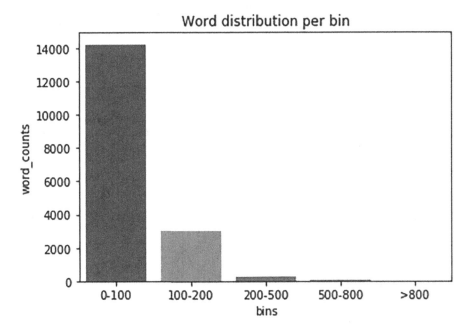

Figure 6-6. *Output*

Most of the descriptions have fewer than 200 words. And more than 80% contains fewer than 100 words.

Text Preprocessing

You already observed the value that text preprocessing lends to many text-related tasks. Let's jump into implementation.

The following is data before text preprocessing.

```
Prod_cat_data['description'][4]
```

The following is the output.

'Key Features of dilli bazaaar Bellies, Corporate Casuals, Casuals
Material: Fabric Occasion: Ethnic, Casual, Party, Formal Color: Pink Heel
Height: 0,Specifications of dilli bazaaar Bellies, Corporate Casuals,
Casuals General Occasion Ethnic, Casual, Party, Formal Ideal For Women Shoe
Details Weight 200 g (per single Shoe) - Weight of the product may vary
depending on size. Heel Height 0 inch Outer Material Fabric Color Pink'

```
# Remove punctuation
Prod_cat_data['description'] = Prod_cat_data['description']
.str.replace(r'[^\w\d\s]', ' ')
# Replace whitespace between terms with a single space
Prod_cat_data['description'] = Prod_cat_data['description']
.str.replace(r'\s+', ' ')
# Remove leading and trailing whitespace
Prod_cat_data['description'] = Prod_cat_data['description']
.str.replace(r'^\s+|\s+?$', '')
# converting to lower case
Prod_cat_data['description'] = Prod_cat_data['description'].str.lower()
# Replace numbers like price values with 'numbr'
Prod_cat_data['description'] = Prod_cat_data['description']
.str.replace(r'\d+(\.\d+)?', 'numbr')
```

```
Prod_cat_data['description'][4]
```

The following is the output.

'key features of dilli bazaaar bellies corporate casuals casuals material
fabric occasion ethnic casual party formal color pink heel height numbr
specifications of dilli bazaaar bellies corporate casuals casuals general

occasion ethnic casual party formal ideal for women shoe details weight numbr g per single shoe weight of the product may vary depending on size heel height numbr inch outer material fabric color pink'

The stop words are imported from the NLTK library and are removed from the description. There are two kinds of stop words.

- General stop words like *an* and *in* appear everywhere irrespective of the domain. It doesn't matter where the text is coming from.

- There are few domain-specific stop words. For example, *buy, com,* and *cash* can appear only in certain domains, such as e-commerce and retail. We need to remove them as well.

```
# Removing Stopwords
stop = stopwords.words('english')
pattern = r'\b(?:{})\b'.format('|'.join(stop))
Prod_cat_data['description'] = Prod_cat_data['description']
.str.replace(pattern, '')
Prod_cat_data['description'] = Prod_cat_data['description']
.str.replace(r'\s+', ' ')# Removing single characters
Prod_cat_data['description'] = Prod_cat_data['description']
.apply(lambda a: " ".join(a for a in a.split() if len(a)>1))

Prod_cat_data['description'][4]
```

The following is the output.

'key features dilli bazaaar bellies corporate casuals casuals material fabric occasion ethnic casual party formal color pink heel height numbr specifications dilli bazaaar bellies corporate casuals casuals general occasion ethnic casual party formal ideal women shoe details weight numbr per single shoe weight product may vary depending size heel height numbr inch outer material fabric color pink'

Let's plot word cloud on descriptions to get to know the words that are appearing the greatest number of times.

```
wordcloud = WordCloud(background_color="white", width = 800, height = 400).
generate(' '.join(Prod_cat_data['description']))
```

```
plt.figure(figsize=(15,8))
plt.imshow(wordcloud)
plt.axis("off")
plt.show()
```

Figure 6-7 shows the descriptions word cloud, which features the frequently appearing words.

Figure 6-7. *Output*

There are a lot of domain-related words occurring in the corpus, which add no value to the task. For example, the word *rs* is an Indian currency present in most documents but not useful. Let's remove these domain-related stop words and plot again.

```
# Removing domain-related stop words from the description

specific_stop_words = ["numbr", "rs","flipkart","buy","com","free","day","c
ash","replacement","guarantee","genuine","key","feature","delivery","produc
ts","product","shipping", "online","india","shop"]
Prod_cat_data['description'] = Prod_cat_data['description'].apply(lambda a:
" ".join(a for a in a.split() if a not in specific_stop_words))
```

The following is the word cloud after removing domain-related stop words.

```
#Wordcloud after removing domain related stop words
```

```
wordcloud = WordCloud(background_color="white", width = 800, height = 400).
generate(' '.join(Prod_cat_data['description']))
plt.figure(figsize=(15,8))
plt.imshow(wordcloud)
plt.axis("off")
plt.show()
```

Figure 6-8 shows the word cloud after removing domain-related stop words.

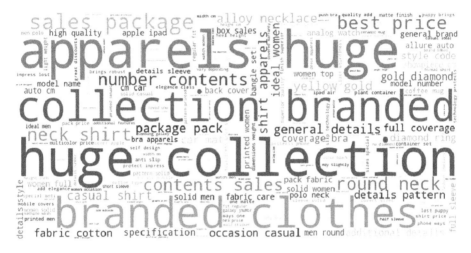

Figure 6-8. *Output*

Note that there are words related to the product categories.

Once the data is cleaned, we can proceed with the feature engineering.

Feature Engineering

To build the text classification model, we need to first convert the text data into features. For a detailed explanation of feature engineering, please refer to Chapter 1. We are considering only one feature described in this project: extracted words from the description after text preprocessing and using them as the vocabulary.

We use deep learning algorithms to build the classifier, and feature extraction should be carried out accordingly. We use the Keras tokenizer function to generate features. We are setting max_length to 200, which means we are only considering 200 features for the classifier. This number also decides the accuracy, and the ideal number can be obtained from hyperparameter tuning.

```
MAX_LENGTH= 200
prod_tok = Tokenizer()
prod_tok.fit_on_texts(Prod_cat_data['description'])
clean_description = prod_tok.texts_to_sequences(Prod_cat_
data['description'])
#padding
X = pad_sequences( clean_description, maxlen= max_length)
```

Like features, we need to also convert the target variable. We use label encoding to do so. The function used is LabelEncoder from sklearn.

```
# Label encoder for Target variable
from sklearn.preprocessing import LabelEncoder
num_class = len(np.unique(Prod_cat_data.product_category_tree.values))
y = Prod_cat_data['product_category_tree'].values
encoder = LabelEncoder()
encoder.fit(y)
y = encoder.transform(y)
```

Train-Test Split

The data is split into two parts: one for training the model and one for evaluating the model.

The train_test_split library from sklearn.model_selection is imported to split the data frame.

```
#train test split
from sklearn.model_selection import train_test_split

independent_features_build, independent_features_valid, depentent_feature_
build, depentent_feature_valid = train_test_split(X, y, test_size=0.2,
random_state=1) #train 80, test 20
print(independent_features_build.shape)
print(independent_features_valid.shape)
print(depentent_feature_build.shape)
```

```
print(depentent_feature_valid.shape)
```

The following is the output.

```
(14025, 200)
(3507, 200)
(14025,)
(3507,)
```

Model Building

The following list of classifiers is used for creating various classification models, which can be further used for prediction.

- Simple baseline artificial neural networks (ANNs)

- Recurrent neural networks (RNN-LSTM)

- Convolutional neural networks

ANN

Let's start with the basic neural network using imbalanced data. You studied neural network functionalities in Chapter 1. We implement the same for this multiclass problem. We use both imbalanced and balanced data and see which one performs better.

The following is the architecture for the neural network. The input neurons are the max_length defined earlier. Then there is an embedding or hidden layer with a linear activation function by default, but we can also use ReLU. In the end, there is a softmax layer with 14 neurons since there are 14 categories. We are using *rmsprop* optimizer with categorical cross-entropy as a loss function.

```
model_inp = Input(shape=(MAX_LENGTH, ))
object_layer = Embedding(vocab_size,100,input_length=MAX_LENGTH)(model_inp)
a = Flatten()(object_layer)
a = Dense(30)(a)
# default activation function is linear, we can make use of relu.
model_pred = Dense(num_class, activation='softmax')(a)
output = Model(inputs=[model_inp], outputs=model_pred)
```

```
output.compile(optimizer='rmsprop',
          loss='categorical_crossentropy',
          metrics=['acc'])

output.summary()
```

Figure 6-9 shows the output of ANN.

```
Layer (type)                    Output Shape                Param #
=================================================================
input_1 (InputLayer)            (None, 200)                 0
_____
embedding_1 (Embedding)         (None, 200, 128)            2286336
_____
flatten_1 (Flatten)             (None, 25600)               0
_____
dense_1 (Dense)                 (None, 32)                  819232
_____
dense_2 (Dense)                 (None, 14)                  462
=================================================================
Total params: 3,106,030
Trainable params: 3,106,030
Non-trainable params: 0
```

Figure 6-9. *Output*

```
filepath="output_ANN.hdf5"
x = ModelCheckpoint(filepath, monitor='val_acc', verbose=1, save_best_
only=True, mode='max')

#fir the model
out = output.fit([indepentent_features_build], batch_size=64,
y=to_categorical(depentent_feature_build), verbose=1, validation_
split=0.25,
          shuffle=True, epochs=5, callbacks=[x])

#predict
output_pred = output.predict(indepentent_features_valid)
output_pred = np.argmax(output_pred, axis=1)
accuracy_score(depentent_feature_valid, output_pred)
```

The following is the output.

```
0.974337040205303
```

The network has achieved 97.4% of accuracy with the network for imbalanced data. Let's try the same network with balanced data and see if there are any improvements in the results.

Long Short-Term Memory: Recurrent Neural Networks

We tried out normal neural networks which don't capture the sequence of the data. Let's try a long short-term memory network where the sequence is also captured while training the model. This is best suited for text since text data is sequential.

The following is the architecture for the neural network. The input neurons are the max_length we defined earlier. An LSTM layer follows an embedding layer. In the end, there is a softmax layer with 14 neurons. We are using a *rmsprop* optimizer with categorical cross-entropy as a loss function.

```
model_inp = Input(shape=(MAX_LENGTH, ))

#define embedding layer
object_layer = Embedding(vocab_size,100,input_length=MAX_LENGTH)(model_inp)

#add LSTM layer
a = LSTM(60)(object_layer)

#add dense layer
a = Dense(30)(a) #default activation function is linear, we can make use of
relu.

#final
model_pred = Dense(num_class, activation='softmax')(a)
output = Model(inputs=[model_inp], outputs=model_pred)

#compile
output.compile(optimizer='rmsprop',loss='categorical_
crossentropy',metrics=['acc'])
output.summary()
```

Figure 6-10 shows the output of LSTM.

```
Layer (type)                Output Shape            Param #
=================================================================
input_4 (InputLayer)        (None, 200)             0

embedding_4 (Embedding)     (None, 200, 128)        2286336

lstm_1 (LSTM)               (None, 64)              49408

dense_7 (Dense)             (None, 32)              2080

dense_8 (Dense)             (None, 14)              462
=================================================================
Total params: 2,338,286
Trainable params: 2,338,286
Non-trainable params: 0
```

Figure 6-10. *Output*

```
filepath="output_LSTM.hdf5"

#model checkpoint
x = ModelCheckpoint(filepath, monitor='val_acc', verbose=1, save_best_
only=True, mode='max')

#fitting
out = output.fit([indepentent_features_build], batch_size=64, y=to_
categorical(depentent_feature_build), verbose=1, validation_split=0.25,
         shuffle=True, epochs=5, callbacks=[x])

output.load_weights('output_LSTM.hdf5')


#predicting on validation data
output_pred = output.predict(indepentent_features_valid)
output_pred = np.argmax(output_pred, axis=1)

#score
accuracy_score(depentent_feature_valid, output_pred)
```

The following is the output.

```
0.9714856002281153
```

The network has achieved 97.1% of accuracy with the LSTM network for imbalanced data. Let's see how the accuracy and validation accuracy vary as we increase epochs.

```
dfaccuracy = pd.DataFrame({'Number of epoch':out.epoch, 'Model hist': out.
history['acc'], 'Model Perd': out.history['val_acc']})
```

```
#line for train accuracy
g = sns.pointplot(x="Number of epoch", y="Model hist", data=dfaccuracy,
fit_reg=False)
```

```
#line for test
g = sns.pointplot(x="Number of epoch", y="Model Perd", data=dfaccuracy,
fit_reg=False, color='Red')
```

Figure 6-11 shows how the accuracy and validation accuracy is varying as we increase epochs for LSTM.

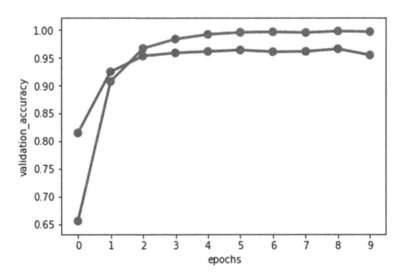

Figure 6-11. *Output*

Convolutional Neural Networks

The last type of network we build is a convolutional neural network. These types of networks are mainly used for image processing. But recent trends show that CNN is performing well on text data if the right parameters are used. Let's see how well it performs in this case.

The following is the architecture for the neural network. The input neurons are the max_length defined earlier. Everything remains the same. Just as there is are convolution and max-pooling layers between an embedding layer and LSTM layer, there is a softmax layer with 14 neurons. We are using a *rmsprop* optimizer with categorical cross-entropy as a loss function.

```
model_inp = Input(shape=(MAX_LENGTH, ))

# define the layer
object_layer = Embedding(vocab_size,100,input_length=MAX_LENGTH)(model_inp)

#conv layer
a = Conv1D(60, 10)(object_layer) #default activation function is linear, we
can make use of relu.

#add pooling layer
a = MaxPooling1D(pool_size=2)(a)
#add LSTM
a = LSTM(60)(a)
a = Dense(30)(a)

#final layer
model_pred = Dense(num_class, activation='softmax')(a)
output = Model(inputs=[model_inp], outputs=model_pred)

#compile
output.compile(optimizer='rmsprop',loss='categorical_
crossentropy',metrics=['acc'])
output.summary()
```

Figure 6-12 shows the output of CNN.

```
Layer (type)                    Output Shape              Param #
=================================================================
input_5 (InputLayer)            (None, 200)               0
_____
embedding_5 (Embedding)         (None, 200, 128)          2286336
_____
conv1d_1 (Conv1D)               (None, 196, 64)           41024
_____
max_pooling1d_1 (MaxPooling1    (None, 49, 64)            0
_____
lstm_2 (LSTM)                   (None, 64)                33024
_____
dense_9 (Dense)                 (None, 32)                2080
_____
dense_10 (Dense)                (None, 14)                462
=================================================================
Total params: 2,362,926
Trainable params: 2,362,926
Non-trainable params: 0
_____
```

Figure 6-12. *Output*

```
filepath="output_CNN.hdf5"
x = ModelCheckpoint(filepath, monitor='val_acc', verbose=1, save_best_
only=True, mode='max')
out = output.fit([indepentent_features_build], batch_size=64, y=to_
categorical(depentent_feature_build), verbose=1, validation_split=0.25,
        shuffle=True, epochs=5, callbacks=[x])

output.load_weights('output_CNN.hdf5')
predicted = output.predict(indepentent_features_valid)

predicted = np.argmax(predicted, axis=1)
accuracy_score(depentent_feature_valid, predicted)
```

The following is the output.

```
0.9663530082691759
```

The network has achieved 96.6% of accuracy with the CNN network for imbalanced data. Let's look at how the accuracy and validation accuracy vary as we increase epochs.

```
dfaccuracy = pd.DataFrame({'epochs':out.epoch, 'accuracy': out.
history['acc'], 'validation_accuracy': out.history['val_acc']})
```

```
#plot
g = sns.pointplot(x="epochs", y="accuracy", data=dfaccuracy, fit_reg=False)
```

```
#plot test accuracy
g = sns.pointplot(x="epochs", y="validation_accuracy", data=dfaccuracy,
fit_reg=False, color='green')
```

Figure 6-13 shows how the accuracy and validation accuracy is varying as we increase epochs for CNN.

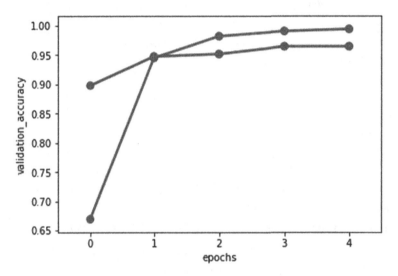

Figure 6-13. *Output*

Evaluating the Model

To evaluate the performance of the classifiers, *k-fold cross validation* is widely used.

Here the data is partitioned into *k* equally sized folds. For each *k* iteration, the training is performed on *k*-1 folds, and the remaining fold evaluates the model. Each iteration error is used for the computation of the overall error estimate.

The advantage of this method is that each observation appears at least once in the training data and at least once in the testing data.

Hyperparameter Tuning

A *randomized search* is an approach to parameter tuning that builds and evaluates a model for each combination of algorithm parameters specified in a grid.

We applied hyperparameter tuning on the CNN algorithm using a Keras classifier and randomized search CV method.

The following are the steps for hyperparameter tuning.

```
def build(num_filters, kernel_size, vocab_size, embedding_dim, maxlength):
    output = Sequential()

    #add embeddings
    output.add(layers.Embedding(vocab_size, embedding_dim, input_
    length=maxlength))
    output.add(layers.Conv1D(num_filters, kernel_size)) #default activation
    function is linear, we can make use of relu.
    output.add(layers.GlobalMaxPooling1D())
    output.add(layers.Dense(20))
#default activation function is linear, we can make use of relu.
    output.add(layers.Dense(num_class, activation='softmax'))
        output.compile(optimizer='rmsprop',loss='categorical_crossentropy
        ',metrics=['accuracy'])
    return output
```

Run the randomized search.

```
from keras.wrappers.scikit_learn import KerasClassifier
from sklearn.model_selection import RandomizedSearchCV

#build the models with parameters
output = KerasClassifier(build_fn=build,
                              epochs=5, batch_size=64,
                              verbose=False)
out = RandomizedSearchCV(estimator=output, param_distributions={'num_
filters': [30, 60, 100],'kernel_size': [4, 6, 8],'embedding_dim':
[40],'vocab_size': [17862],
  'maxlength': [180]},cv=4, verbose=1, n_iter=5)
out_result = out.fit(indepentent_features_build, depentent_feature_build)
```

Find the best parameters of the model.

```
# Evaluate testing set
```

```
test_accuracy = out.score(indepentent_features_valid, depentent_feature_
valid)
print(out_result.best_params_)
```

The following is the output.

```
{'vocab_size': 16553, 'num_filters': 128, 'maxlength': 200, 'kernel_size':
5, 'embedding_dim': 50}
```

```
print(out_result.best_score_)
```

The following is the output.

```
0.980473461948488
```

After running the randomized search CV, we found out that the model with the following parameters has the best accuracy, 98.04.

Results

The accuracy measure evaluates the performance of a model. We built multiple types of neural networks starting from ANN. Later, we used LSTM networks and, in the end, CNN. Let's compare all the accuracy and select the model with the best accuracy. Table 6-3 shows all the results in one place.

Table 6-3. *Output Summary*

Classifier	Accuracy
Simple Baseline ANN	0.97434
RNN-LSTM	0.97149
CNN with 1Dimensional	0.96635
CNN with 1Dimensional after Hyper tuning	**0.98047**

The CNN model with 1D gave better performance compared to other models with the following parameters.

- The vocab size is 16553.

- The number of filters of CNN is 128.

- The maximum input length or features is 200.

- The kernel size is 5.

- The embedding layer dimension is 50.

Summary

This chapter implemented multiclass classification using different deep learning algorithms.

You learned from the implemented projects that 1D convolutional neural networks perform better than the simple baseline ANN model. You can try with pretrained embeddings like word2vec and GloVe for more accurate results. You can apply more text preprocessing steps like lemmatization, spelling correction, and so on. The chapter used the "description" feature only, but you can also use product specifications to evaluate if it adds more value.

## CHAPTER 7

Predicting Duplicate Questions in Quora

Quora is a question-and-answer platform for folks to connect and gain knowledge on various topics. Anyone can ask a question, and anyone can write answers to these questions, thereby bringing variety in views.

Problem Statement

Questions flow in each day, and thousands are answering them. There is a very good probability that questions are the same literally or with similar intent, which creates unnecessary duplicate questions and answers in the system. This can be data-heavy and the user experience takes a hit.

For instance, "Who is the best cricketer in the world now?" and "Currently, which cricketer ranks top in the world?" They are the same question but asked differently. So, identifying whether given questions are similar or not is the main goal of the problem. If we can solve this problem with the help of NLP, that reduces significant duplication and enhances the user experience.

Approach

This is a straightforward sentence comparison NLP task. We use both unsupervised and supervised approaches to solve this problem.

© Akshay Kulkarni, Adarsha Shivananda and Anoosh Kulkarni 2022
A. Kulkarni et al., *Natural Language Processing Projects*, https://doi.org/10.1007/978-1-4842-7386-9_7

Unsupervised Learning

Let's use various pretrained models like Sentence-BERT, GPT, and doc2vec to solve this problem. Figure 7-1 is a flow chart of the process of unsupervised learning.

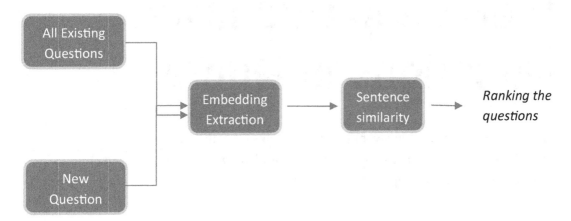

Figure 7-1. *Unsupervised learning flow*

The following are the steps for unsupervised learning.

1. Stack the question 1 and question 2 columns vertically to get embeddings of all questions.

2. Encode questions using various sentence embeddings techniques like Sentence-BERT, OpenAI GPT, and doc2vec.

3. Find similar questions using cosine similarity between the input vector and the original data set, and return the top N questions with the highest cosine similarity score.

Supervised Learning

In supervised learning, we are training the deep learning classifier to predict whether two sets of questions are similar or not. The output is a probabilistic score between two questions. Figure 7-2 illustrates the flow of the process.

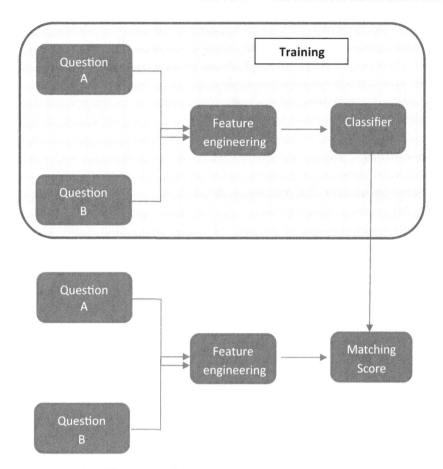

Figure 7-2. Supervised learning flow

Data set

We are using the Quora Question Pairs data set from Kaggle. You can download the data from www.kaggle.com/c/quora-question-pairs/data.

The data contains IDs and questions. It also contains a label to implicate whether two questions are duplicates. If it's duplicate then the value will be 1, otherwise 0.

Figure 7-3 shows the output.

	id	qid1	qid2	question1	question2	is_duplicate
0	0	1	2	What is the step by step guide to invest in sh...	What is the step by step guide to invest in sh...	0
1	1	3	4	What is the story of Kohinoor (Koh-i-Noor) Dia...	What would happen if the Indian government sto...	0
2	2	5	6	How can I increase the speed of my internet co...	How can Internet speed be increased by hacking...	0
3	3	7	8	Why am I mentally very lonely? How can I solve...	Find the remainder when [math]23^{24}[/math] i...	0
4	4	9	10	Which one dissolve in water quikly sugar, salt...	Which fish would survive in salt water?	0

Figure 7-3. *Output*

Now that we understand the problem statement, let's jump into implementation.

Implementation: Unsupervised Learning

Data Preparation

Let's import the necessary libraries.

```
import pandas as pd
import numpy as np
import os

import scipy
import string
import csv

#import nltk
import nltk
nltk.download('stopwords')
nltk.download('punkt')
nltk.download('wordnet')

#immport tokenize, stopwords
from nltk.tokenize import word_tokenize
from nltk.corpus import stopwords
from nltk.stem import WordNetLemmatizer

#import warnings
import warnings
```

```
#import sklearn and matplotlib
from sklearn import preprocessing
import spacy
import matplotlib.pyplot as plt
import plotly.graph_objects as go

#import warning
warnings.filterwarnings('ignore')
import re

#import the data
train=pd.read_csv('quora_train.csv')
train1=train.copy()

train.head()
```

Figure 7-4 shows the output.

	id	qid1	qid2	question1	question2	is_duplicate
0	0	1	2	What is the step by step guide to invest in sh...	What is the step by step guide to invest in sh...	0
1	1	3	4	What is the story of Kohinoor (Koh-i-Noor) Dia...	What would happen if the Indian government sto...	0
2	2	5	6	How can I increase the speed of my internet co...	How can Internet speed be increased by hacking...	0
3	3	7	8	Why am I mentally very lonely? How can I solve...	Find the remainder when [math]23^{24}[/math] i...	0
4	4	9	10	Which one dissolve in water quikly sugar, salt...	Which fish would survive in salt water?	0

Figure 7-4. *Output*

```
#append the both set of questions in dataset
Q1=train1.iloc[:,[2,4]]
Q2=train1.iloc[:,[1,3]]

df = pd.DataFrame( np.concatenate( (Q2.values, Q1.values), axis=0 ) )
df.columns = ['id','question' ]
df
```

Figure 7-5 shows the output.

	id	question
0	1	What is the step by step guide to invest in sh...
1	3	What is the story of Kohinoor (Koh-i-Noor) Dia...
2	5	How can I increase the speed of my internet co...
3	7	Why am I mentally very lonely? How can I solve...
4	9	Which one dissolve in water quikly sugar, salt...
...	...	...
808575	379845	How many keywords are there in PERL Programmin...
808576	155606	Is it true that there is life after death?
808577	537929	What's this coin?
808578	537931	I am having little hairfall problem but I want...
808579	537933	What is it like to have sex with your cousin?

Figure 7-5. *Output*

A. Building Vectors Using doc2vec

doc2vec is similar to word2vec. In word2vec, there is a vector of each word or token. But if we need a vector for a sentence or document, the average of all the word vectors is calculated. In that process, the document might lose some information or context.

Here we evolve to doc2vec, where every document is passed to the model as unique input and vectors are found.

Let's import the necessary libraries and start building the model.

```
# importing doc2vec from gensim
from gensim.models.doc2vec import Doc2Vec, TaggedDocument
```

```
# tokenizing the sentences
tok_quora=[word_tokenize(wrd) for wrd in df.question]
```

```
#creating training data
Quora_training_data=[TaggedDocument(d, [i]) for i, d in
enumerate(tok_quora)]
```

The following is the output.

```
[TaggedDocument(words=['What', 'is', 'the', 'step', 'by', 'step', 'guide',
'to', 'invest', 'in', 'share', 'market', 'in', 'india', '?'], tags=[0]),
```

```
 TaggedDocument(words=['What', 'is', 'the', 'story', 'of', 'Kohinoor', '(',
'Koh-i-Noor', ')', 'Diamond', '?'], tags=[1]),
 TaggedDocument(words=['How', 'can', 'I', 'increase', 'the', 'speed',
'of', 'my', 'internet', 'connection', 'while', 'using', 'a', 'VPN', '?'],
tags=[2]),……
```

```
# trainin doc2vec model
doc_model = Doc2Vec(Quora_training_data, vector_size = 100, window = 5,
min_count = 3, epochs = 25)
```

Let's build a function to get embedding vectors of each sentence. Also, let's make sure we are using only those words from sentences in vocabulary.

```
#function to get vectors from model

def fetch_embeddings(model,tokens):
  tokens = [x for x in word_tokenize(tokens) if x in list(doc_model.
  wv.vocab)]
  #if words is not present then vector becomes zero
  if len(tokens)>=1:
    return doc_model.infer_vector(tokens)
  else:
    return np.array([0]*100)
```

```
#Storing all embedded sentence vectors in a list
```

```
#defining empty list and iterating through all the questions
```

```
doc_embeddings=[]
for w in df.question:
    doc_embeddings.append(list(fetch_embeddings(doc_model, w)))
#conveting it into array
doc_embeddings=np.asarray(doc_embeddings)
```

```
#shape
Shape=(10000,100)
```

```
#Output:
```

Figure 7-6 shows the output.

```
array([[-0.07908138,  0.34195665, -0.3269907 , ..., -0.00650925,
        -0.05626418,  0.717077  ],
       [ 0.03356582,  0.12112963, -0.02335722, ..., -0.29730332,
         0.05184059,  0.3093264 ],
       [-0.10729031,  0.21547607, -0.8347038 , ..., -0.155304  ,
         0.07417022, -0.33322385],
       ...,
       [ 0.00395333,  0.06456551, -0.42053157, ...,  0.09512904,
         0.22979783,  0.13475512],|
       [-0.09726299,  0.2122679 ,  0.0308287 , ...,  0.14147176,
        -0.08733932,  0.35432833],
       [ 0.14448403,  0.21737853, -0.28049508, ...,  0.25516173,
        -0.319338  ,  0.01880774]], dtype=float32)
```

Figure 7-6. *Output*

B. Sentence Transformers Using the BERT Model

Sentence-BERT is the improvised version of BERT for sentences. In this new model, two sentences are passed to get the embeddings and build a layer on top of it. It uses Siamese Networks, in the end, to find the similarity.

Figure 7-7 shows that two sentences are passed via BERT and pooling layers, and similarity is found at the end.

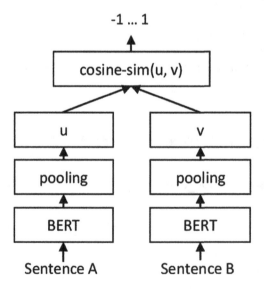

Figure 7-7. *SBERT architecture from https.//arxiv.org/pdf/1908.10084.pdf*

So, let's implement SBERT.

```
#install SBERT
!pip install sentence-transformers
```

```
#import the SBERT
from sentence_transformers import SentenceTransformer
```

```
#let use paraphrase-MiniLM-L12-v2 pre trained model
sbert_model = SentenceTransformer('paraphrase-MiniLM-L12-v2')
```

```
x=[i for i in df.question]
#lets get embeddings for each question
sentence_embeddings_BERT= sbert_model.encode(x)
```

```
#lets see the shape
sentence_embeddings_BERT.shape
(10000, 384)
```

The shape of the embeddings is (10000, 384) because it produces 384-dimensional embeddings. The following is the output.

```
sentence_embeddings_BERT
```

Figure 7-8 shows the output.

```
array([[-0.15299696,  -0.30485195,   0.00183832,  ..., -
0.27034327,
         -0.4260835 ,   0.31928647],
       [-0.16776392,   0.67119426,  -0.51778895,  ..., -
0.08420195,
          0.00470432,   0.3200466 ],
       [ 0.09468909,  -0.00629827,   0.06894321,  ..., -
0.18716985,
         -0.2556718 ,   0.04188535],
       ...,
       [-0.03576591,   0.24553823,   0.10434522,  ...,   0.1781153
,
         -0.00719658,  -0.19804102],
       [ 0.15157805,  -0.32189795,   0.01304498,  ..., -
0.05349696,
         -0.33300248,   0.18506213],
       [ 0.2177582 ,  -0.07103815,   0.11279581,  ...,  -0.5100467
,
          0.05511575,  -0.1291719 ]], dtype=float32)
```

Figure 7-8. *Output*

C. GPT

The GPT (Generative Pretrained Transformer) model is from Open AI. It is one
of the very effective language models which can again perform various tasks like
summarization and Q&A systems.

Figure 7-9 shows GPT architecture.

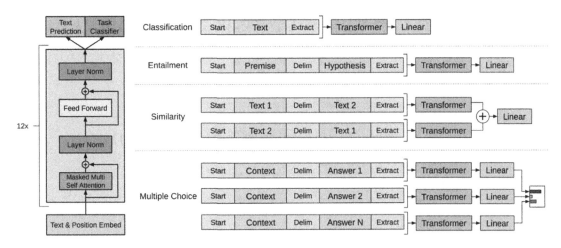

Figure 7-9. *GPT architecture*

```
#Installing the GPT
!pip install pytorch_pretrained_bert

#Importing required tokenizer, OpenAiGPT model
import torch
from pytorch_pretrained_bert import OpenAIGPTTokenizer, OpenAIGPTModel

#initializing the tokenizer
tok_gpt= OpenAIGPTTokenizer.from_pretrained('openai-gpt')

#Initializing the gpt Model
model_gpt= OpenAIGPTModel.from_pretrained('openai-gpt')
model_gpt.eval()
```

Now that we have imported the gpt model let's write a function to get embedding for all the questions.

```
def Fetch_gpt_vectors(question):

    #tokenize words
    words = word_tokenize(question)
    emb = np.zeros((1,768))

    #get vectore for each word
    for word in words:
        w= tok_gpt.tokenize(word)
```

```
        indexed_words = tok_gpt.convert_tokens_to_ids(w)
        tns_word = torch.tensor([indexed_words])

        with torch.no_grad():
            try:
    #get mean vector
                emb += np.array(torch.mean(model_gpt(tns_word),1))
            except Exception as e:
                continue
  emb /= len(words)
  return emb
```

We have the function. Let's use it for the question data set and save vectors in a given array of size 1000, 768.

```
gpt_emb = np.zeros((1000, 768))

# get vectors

for v in range(1000):
    txt = df.loc[v,'question']

    gpt_emb[v] = Fetch_gpt_vectors(txt)

gpt_emb
```

Figure 7-10 shows the output.

```
array([[ 0.12315898,   0.14081994, -0.06503178, ...,
0.25522971,
        -0.29333039,   0.13614352],
       [ 0.29858619,   0.10205475,  0.13587548, ...,
0.06204159,
        -0.12169133,   0.01781069],
       [ 0.12268556,   0.04076901,  0.03162696, ...,
0.16541341,
        -0.29518041,   0.06638591],
        ...,
       [ 0.18672558,   0.00395902,  0.0360269 , ...,   0.2380983
,
        -0.33544575, -0.05039981],
       [ 0.21165186,   0.1225166 ,  0.10084978, ...,
0.12687551,
        -0.27586632,   0.13541676],
       [ 0.15088044, -0.01128645,  0.14384876, ...,
0.28454954,
        -0.30602185,   0.0567844 ]])
```

Figure 7-10. *Output*

Finding Similar questions

Now that there are various types of embeddings, it's time to find similar questions using a similarity measure.

Let's build a function to get embeddings of input vectors. We need to input sentences and models, and it returns embeddings.

```
#defining function to derive cosine similarity

#import
from sklearn.metrics.pairwise import cosine_similarity
from numpy import dot
from numpy.linalg import norm

def cosine_similarity(vec1,vec2):
    #find the score
    return dot(vec1, vec2)/(norm(vec1)*norm(vec2))
```

In this function, we are giving inputs as follows.

```
User: input query from the user
Embeddings: Embeddings from which we have to find similar questions
df: data set from which we are recommending questions

#Function which gets Top N similar questions from data

def top_n_questions(user,embeddings,df):

    #getting cosine similarities of overall data set with input queries
    from user
    x=cosine_similarity(user,embeddings).tolist()[0]
    temp_list=list(x)

    #sorting
    sort_res = sorted(range(len(x)), key = lambda sub: x[sub])[:]
    sim_score=[temp_list[i] for i in reversed(sort_res)]

    #print
    print(sort_res[0:5])

    #index fetching
    L=[]
    for i in reversed(sort_res):
        L.append(i)

    #get the index from dataframe
    return df.iloc[L[0:5], [0,1]]

#function to fetch the results based on the model selected

def get_input_vector(query,model):

    print(query)

    #Doc2vec model
    if model=='Doc2Vec':
      k=fetch_embeddings(doc_model,query)
      k=k.reshape(1, -1)

    # sbert  model
```

```
elif model=='BERT':
  k=sbert_model.encode(str(query))
  k=k.reshape(1, -1)

# gpt model
elif model=='GPT':
  k=Fetch_gpt_vectors(query)

return k
```

All the functions are in place. Let's look at how the results look like.

```
# Example 1 - Doc2vec model


top_n_questions(get_input_vector('How is Narendra Modi as a
person?','Doc2Vec'),doc_embeddings,df)
```

Figure 7-11 shows the output.

```
How is Narendra Modi as a person?
[0.9729089736938477, 0.8310148119926453, 0.8260904550552368, 0.8139867782592773, 0.7872357368469238]
               id                           question

      2166    4310           How is Narendra Modi as a person?

      7404    14467                   How is Tim Cook as a CEO?

      592     1182                    How is time travel possible?

      2211    4399    How is black money stored in a Swiss bank?

      4926    9719        How do I know whether a person is lying?
```

Figure 7-11. *Output*

```
# Example 2 - GPT model
```

```
top_n_questions(get_input_vector('How is Narendra Modi as a
person?','GPT'),gpt_emb,df)
```

Figure 7-12 shows the output.

```
How is Narendra Modi as a person?
[0.9037639668695364, 0.9018443190478334, 0.8992568944325751, 0.8962871094823177, 0.8818656399481746]
```

	id	question
51	103	Will a Blu Ray play on a regular DVD player? I...
23	47	How much is 30 kV in HP?
97	195	Why did harry become a horcrux?
71	143	What is a narcissistic personality disorder?
7	15	How can I be a good geologist?

Figure 7-12. *Output*

Example 3 - BERT

```
top_n_questions(get_input_vector('How is Narendra Modi as a
person?','BERT'),sentence_embeddings_BERT,df)
```

Figure 7-13 shows the output.

```
How is Narendra Modi as a person?
[1.0, 0.8795616626739502, 0.8193296790122986, 0.8173707723617554, 0.8137117624282837]
```

	id	question
2166	4310	How is Narendra Modi as a person?
5998	11766	How can I meet Narendra Modi?
3120	6186	What is the life history of Sundar Pichai?
3628	2267	How can I meet Modi?
1137	2266	How can I get a chance to meet Mr. Narendra Modi?

Figure 7-13. *Output*

As you can see, SBERT gives better results than the other models.

There are always limitations in unsupervised learning, and we can do only so much to increase the accuracy. Let's look at the supervised learning approach to produce the results.

Implementation: Supervised Learning

Let's implement the same problem using supervised learning. In the data, there are two questions and a target variable saying these questions are similar or not. We can use this data and build a text classifier.

Let's get started.

Understanding the Data

```
# import packages required.
import pandas as pd
import numpy as np
import scipy
import os
import string
import csv

#import nltk
import nltk
nltk.download('stopwords')
nltk.download('punkt')
nltk.download('wordnet')

#import tokenizer
from nltk.tokenize import word_tokenize
from nltk.stem import WordNetLemmatizer

#import warnings
import warnings

#import sklearn and matplotlib
from sklearn import preprocessing
import spacy
import matplotlib.pyplot as plt
import plotly.graph_objects as go
```

```
#import warning
warnings.filterwarnings('ignore')
import re

from string import punctuation
from nltk.stem import SnowballStemmer
from nltk.corpus import stopwords
stop_words = set(stopwords.words('english'))

#import Tokenizer from keras
from keras.preprocessing.text import Tokenizer
from keras.preprocessing import sequence
from sklearn.model_selection import train_test_split

#importing Keras necessary libraries
from keras.models import Sequential, Model
from keras.layers import Input, Embedding, Dense, Dropout, LSTM
```

Let's import the whole data set which was downloaded earlier.

```
#importing train data - Import the full data
quora_questions=pd.read_csv('Quora.csv')
```

Preprocessing the Data

Let's create a text preprocessing function that can be used later for all columns in the data set as well as new input data from the user.

```
#function for data cleaning
def txt_process(input_text):

    # Removing punctuation from input text
    input_text = ''.join([x for x in input_text if x not in punctuation])

    # Cleaning the text
    input_text = re.sub(r"[^A-Za-z0-9]", " ", input_text)
    input_text = re.sub(r"\'s", " ", input_text)

    # remove stop words
    input_text = input_text.split()
```

```
input_text = [x for x in input_text if not x in stop_words]
input_text = " ".join(input_text)

# Return a list of words
return(input_text)
```

Let's call the data text cleaning function on both question1 and question 2 so that we have clean text.

```
#applying above function on both question ids
quora_questions['question1_cleaned'] = quora_questions.apply(lambda x: txt_
process(x['question1']), axis = 1)
quora_questions['question2_cleaned'] = quora_questions.apply(lambda x: txt_
process(x['question2']), axis = 1)
```

Text to Feature

Let's stack both question IDs together so we can cover all words from both columns. Then we can tokenize these words to convert them to numbers.

```
#stacking
question_text = np.hstack([quora_questions.question1_cleaned, quora_
questions.question2_cleaned])
```

```
#tokenizing
tokenizer = Tokenizer()
tokenizer.fit_on_texts(question_text)
```

```
#creating new columns for both ids where tokenized form of sentence is
created
quora_questions['tokenizer_1'] = tokenizer.texts_to_sequences(quora_
questions.question1_cleaned)
quora_questions['tokenizer_2'] = tokenizer.texts_to_sequences(quora_
questions.question2_cleaned)
```

```
quora_questions.head(5)
```

Figure 7-14 shows the output.

	id	qid1	qid2	question1	question2	is_duplicate	question1_cleaned	question2_cleaned	tokenizer_1	tokenizer_2
0	0	1	2	What is the step by step guide to invest in sh...	What is the step by step guide to invest in sh...	0	What step step guide invest share market india	What step step guide invest share market	[1, 1054, 1054, 3819, 577, 431, 369, 9]	[1, 1054, 1054, 3819, 577, 431, 369]
1	1	3	4	What is the story of Kohinoor (Koh-i-Noor) Dia...	What would happen if the Indian government sto...	0	What story Kohinoor KohiNoor Diamond	What would happen Indian government stole Kohi...	[1, 325, 2313, 2313, 3820]	[1, 14, 132, 42, 133, 4595, 2313, 2313, 3820, ...
2	2	5	6	How can I increase the speed of my internet co...	How can Internet speed be increased by hacking...	0	How I increase speed internet connection using...	How Internet speed increased hacking DNS	[3, 2, 109, 432, 237, 1461, 84, 2960]	[3, 237, 432, 2037, 1319, 8527]
3	3	7	8	Why am I mentally very lonely? How can I solve...	Find the remainder when [math]23^{24} [/math] i...	0	Why I mentally lonely How I solve	Find remainder math2324math divided 2423	[4, 2, 1462, 3821, 3, 2, 578]	[37, 8528, 8529, 8530, 8531]
4	4	9	10	Which one dissolve in water...	Which fish would survive in salt...	0	Which one dissolve water...	Which fish would survive...	[8, 15, 2961, 161, 5948, 1304, 1305,	[8, 4258, 14, 1928,

Figure 7-14. *Output*

```
#combining both tokens in one list question1 followed by question2
quora_questions['tokenizer'] = quora_questions['tokenizer_1'] + quora_
questions['tokenizer_2']

#defining max length
m_len = 500

#max tokens
max_token = np.max(quora_questions.tokenizer.max())
```

Model Building

Let's split the data into test and train and build an LSTM classifier. Along with LSTM, let's use the dropout layer as well to reduce the overfitting. There is the sigmoid layer, given it's a binary classification.

```
#defining X and target data
y = quora_questions[['is_duplicate']]
X = quora_questions[['tokenizer']]

#padding X with a maximum length
X = sequence.pad_sequences(X.tokenizer, maxlen = m_len)
```

```
#splitting data into train and test
X_train,X_test,y_train,y_test=train_test_split(X, y, test_size=0.25,
random_state=10)
```

Let's train the model on training data using the LSTM model.

```
#defining the LSTM model
quora_model = Sequential()
```

```
#adding embeedding layer
quora_model.add(Embedding(70000, 64))
```

```
#adding drop out layer
quora_model.add(Dropout(0.15))
```

```
#LSTM layer
quora_model.add(LSTM(16))
```

```
#adding sigmoid layer
quora_model.add(Dense(1, activation = 'sigmoid'))
```

```
#defining loss and optimizer
quora_model.compile(loss='binary_crossentropy', optimizer='SGD',
metrics=['accuracy'])
```

```
quora_model.summary()
```

Figure 7-15 shows the output.

```
Model: "sequential"

_____
Layer (type)                Output Shape              Param #
=================================================================
embedding (Embedding)       (None, None, 200)         10000000

_____
dropout (Dropout)           (None, None, 200)         0

_____
lstm (LSTM)                 (None, 32)                29824

_____
dense (Dense)               (None, 1)                 33
=================================================================
Total params: 10,029,857
Trainable params: 10,029,857
Non-trainable params: 0
_____
```

Figure 7-15. *LSTM model summary*

```
#training the model and validating on test data
quora_model.fit(X_train, y_train, epochs = 2, batch_size=64,validation_
data=(X_test,y_test))

Epoch 1/2
2527/2527 [==============================] - 2342s 927ms/step - loss:
0.5273 - accuracy: 0.7398 - val_loss: 0.5014 - val_accuracy: 0.7571
Epoch 2/2
2527/2527 [==============================] - 2307s 913ms/step - loss:
0.4874 - accuracy: 0.7670 - val_loss: 0.4781 - val_accuracy: 0.7701
<keras.callbacks.History at 0x7fc1d72b1550>
```

Evaluation

Let's look at how the model performs on train and test data using various parameters like confusion matrix and F1 score.

```
# evaluation of the model
import sklearn
from sklearn.metrics import classification_report

#prediction on train data
tr_prediction=quora_model.predict(X_train)
```

```
#replacing probabilities >0.5 with 1 and other 0
tr_prediction[tr_prediction>0.5]=1
tr_prediction[tr_prediction<0.5]=0
tr_prediction

#true values of train data
tr_true=y_train.values

#accuracy
Accuracy=sklearn.metrics.accuracy_score(np.array(tr_true),
                                np.array(tr_prediction))

print(Accuracy)
0.7811906400332337

#classification report with f1 score

print(classification_report(tr_true, tr_prediction, target_names=['Not
similar','similar']))
```

Figure 7-16 shows the output.

	precision	recall	f1-score	support
Not similar	0.78	0.91	0.84	198868
similar	0.79	0.55	0.65	116475
accuracy			0.78	315343
macro avg	0.78	0.73	0.75	315343
weighted avg	0.78	0.78	0.77	315343

Figure 7-16. *Output*

We are getting an F1 of 78% on training data. Let's see how the model behaves on test data.

```
#predicting on test data
test_prediction=quora_model.predict(X_test)
```

```
#generating classes
test_prediction[test_prediction>0.5]=1
test_prediction[test_prediction<0.5]=0
test_prediction

#true values for test
test_true=y_test.values

# accuracy on test data
Accuracy=sklearn.metrics.accuracy_score(np.array(test_true),
                                np.array(test_prediction))

print('Accuracy is %f'%(Accuracy*100)+' %')
Accuracy is 78.152545 %

print(classification_report(test_true, test_prediction, target_names=['Not
similar','similar']))
```

Figure 7-17 shows the output.

	precision	recall	f1-score	support
Not similar	0.78	0.91	0.84	56156
similar	0.79	0.56	0.65	32788
accuracy			0.78	88944
macro avg	0.78	0.73	0.75	88944
weighted avg	0.78	0.78	0.77	88944

Figure 7-17. *Output*

We get a 78% F1 on test data as well, which is a great sign. This means that the model is not overfitting and accuracy looks promising. With further efforts into fine-tuning and more epochs, we can try to increase the same.

Predictions on New Sentence Pairs

Let's build a function to get the probability score for given pair of questions.

```
def find_similarity_score(q1,q2):
```

```
#clean first question
Q1_C= txt_process(q1)
#print(q1)

#clean first question
Q2_C = txt_process(q2)
#print(q2)

#converting 1st question into tokens
Q1_C = tokenizer.texts_to_sequences([Q1_C])

#converting 2nd question into token
Q2_C = tokenizer.texts_to_sequences([Q2_C])

#combining both tokens as we did for train data
Q_final = Q1_C[0] + Q2_C[0]

#padding combined sequence to max length
Q_Test = sequence.pad_sequences([Q_final], maxlen = 500)

#predicting probability of given pair
Prob=quora_model.predict(Q_Test)
print(Prob)

#if p>0.5 then similar
if Prob[0]>0.5:
  return 'Quora Questions are similar'
else:
  return 'Quora Questions are Not similar'
#example 1
find_similarity_score('Who is Narendra Modi?','What is identity of Narendra
Modi?')

[[0.70106715]]
Quora Questions are similar

#example 2
find_similarity_score('is there life after death?','Do people belive in
afterlife')
```

```
[[0.550213]]
Quora Questions are similar

#example 3
find_similarity_score('Should I have a hair transplant at age 24? How much
would it cost?','How much cost does hair transplant require?')

[[0.17271936]]
Quora Questions are Not similar
```

You can see how good the probabilities are. In the first two examples, it's greater than 0.5, and the questions are similar. Example 3 is the scenario where questions are not similar.

Conclusion

We successfully built both models for finding similar Quora questions, supervised and unsupervised learning. We explored various latest pretrained embeddings to solve this problem. In supervised learning, we trained a deep learning classifier to predict the similarity without using any embeddings.

Next, let's try the following steps to improvise.

1. We can go with the hybrid approach. We can take pretrained embeddings that act as the base vectors and build a classifier on top of it—a transfer learning approach. We explore this concept in later chapters of this book.

2. We can also try different and advanced deep learning architectures like attention mechanism instead of simple LSTM with dropout.

3. Hyperparameter tuning of the parameters in any architecture increases the accuracy.

CHAPTER 8

Named-Entity Recognition Using CRF and BERT

Named-entity recognition (NER) is a natural language processing technique. It is also called *entity identification* or *entity extraction*. It identifies named entities in text and classifies them into predefined categories. For example, extracted entities can be the names of organizations, locations, times, quantities, people, monetary values, and more present in text.

With NER, key information is often extracted to learn what a given text is about, or it is used to gather important information to store in a database.

NER is used in many applications across many domains. NER is extensively used in biomedical data. For instance, it is used for DNA identification, gene identification, and the identification of drug names and disease names. Figure 8-1 shows an example of a medical text-related NER that extracts symptoms, patient type, and dosage.

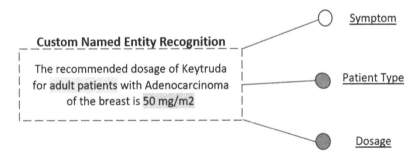

Figure 8-1. *NER on medical text data*

NER is also used for optimizing the search queries and ranking search results. It is sometimes combined with topic identification. NER is also used in machine translation.

A. Kulkarni et al., *Natural Language Processing Projects*, https://doi.org/10.1007/978-1-4842-7386-9_8

There are a lot of pretrained general-purpose libraries that use NER. For example, spaCy—an open source Python library for various NLP tasks. And NLTK (natural language tool kit) has a wrapper for the Stanford NER, which is simpler in many cases.

These libraries only extract a certain type of entities, like name, location, and so on. If you need to extract something very domain-specific, such as the name of a medical treatment, it is impossible. You need to build custom NER in those scenarios. This chapter explains how to build a custom NER model.

Problem Statement

The goal is to extract a named entity from movie trivia. For example, the tags are for movie name, actor name, director name, and movie plot. General libraries might extract the names but don't differentiate between actor and director, and it would be challenging to extract movie plots. We need to build the customer model that predicts these tags for the sentences on movies.

Essentially, we have a data set that talks about the movies. It consists of sentences or questions on movies, and each of those words in the sentence has a predefined tag. We need to build the NER model to predict these tags.

Along the way, we need to understand these concepts.

1. Build the model using various algorithms.

2. Design a metric to measure the performance of the model.

3. Understand where the model fails and what might be the reason for the failure.

4. Fine-tune the model.

5. Repeat these steps until we achieve the best accuracy on the test data.

Methodology and Approach

NER identifies the entities in a text and classifies them into predefined categories such as location, a person's name, organization name, and more. But for this problem, we need to tag the director's name, actor's name, genre, and movie character (likewise, there are 25 such tags defined in the data set) for the entities in a sentence. So NER alone does not suffice. Here let's build custom models and train them using conditional random fields and BERT.

The steps to solve this problem are as follows.

1. Data collection

2. Data understanding

3. Data preprocessing

4. Feature mapping

5. Model Building

   • Conditional random fields

   • BERT

6. Hyperparameter tuning

7. Evaluating the model

8. Prediction on random sentences

Figure 8-2 shows how the product works at a high level.

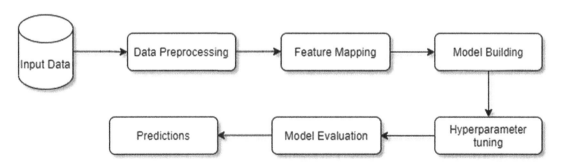

Figure 8-2. *Approach flow*

Implementation

We understood the problem statement as well as various approaches to solve it. Let's start the implementation of the project. We begin with importing and understanding the data.

Data

We used data from the MIT movie corpus, which is in .bio format. Download the trivia10k13train.bio and trivia10k13test.bio data sets from `https://groups.csail.mit.edu/sls/downloads/movie/`.

Now let's convert the data into a pandas data frame using the following code.

```
#create a function to add a column sentence that indicates the sentence
u=id for each txt file as a preprocessing step.

import pandas as pd
def data_conversion(file_name):
    df_eng=pd.read_csv(file_name,delimiter='\t',header=None,skip_blank_
lines=False)
    df_eng.columns=['tag','tokens']
    tempTokens = list(df_eng['tokens'])
    tempSentence = list()
    count = 1
    for i in tempTokens:
        tempSentence.append("Sentence" + str(count))
        if str(i) == 'nan':
            count = count+1
    dfSentence = pd.DataFrame (tempSentence,columns=['Sentence'])
    result = pd.concat([df_eng, dfSentence], axis=1, join='inner')
    return result

#passing the text files to function
trivia_train=data_conversion('trivia10k13train.txt')
trivia_test=data_conversion('trivia10k13test.txt')
```

Train Data Preparation

Let's look at the first five rows of the training data.

```
trivia_train.head()
```

Figure 8-3 shows the output for the first five rows of the training data.

	tag	tokens	Sentence
0	B-Actor	steve	Sentence1
1	I-Actor	mcqueen	Sentence1
2	O	provided	Sentence1
3	O	a	Sentence1
4	B-Plot	thrilling	Sentence1

Figure 8-3. *Sample training data*

Next, we check the number of rows and columns present in the training data set.

```
trivia_train.shape
```

The following is the output.

```
(166638, 3)
```

There are a total of 166,638 rows and three columns. Let's check how many unique words are present in the training data set.

```
trivia_train.tokens.nunique()
```

The following is the output.

```
10986
```

There are 10,986 unique words and 7816 total sentences for the training data. Now, let's check if there are any null values present in the training data set.

```
trivia_train.isnull().sum()
```

The following is the output.

```
tag          7815
tokens       7816
Sentence        0
dtype: int64
```

There are 7816 null rows. Let's drop the null rows using the following code.

```
trivia_train.dropna(inplace=True)
```

Test Data Preparation

Let's look at the first five rows of the test data.

```
trivia_test.head()
```

Next, we check the number of rows and columns present in the test data set.

```
trivia_test.shape
```

The following is the output.

```
(40987, 3)
```

There are a total of 40,987 rows and three columns. Let's check how many unique words are present in the test data set.

```
trivia_test.tokens.nunique()
```

The following is the output.

```
5786
```

There are 5786 unique words and 1953 total sentences for test data. Now, let's check if there are any null values present in the test data set.

```
trivia_test.isnull().sum()
```

The following is the output.

```
tag           1952
tokens        1952
Sentence         0
dtype: int64
```

There are 1952 null rows. Let's drop the null rows using the following code.

```
trivia_test.dropna(inplace=True)
```

The data set consists of three columns after extraction.

- **tag** is the category of words

- **tokens** consist of words

- **sentence** is the sentence number in which a word belongs

There are a total of 24 (excluding the O tag) unique tags in the given training data set. Their distribution is as follows.

```
#below is the code to get the distribution plot for the tags.

trivia_train[trivia_train["tag"]!="O"]["tag"].value_counts().
plot(kind="bar", figsize=(10,5))
```

Figure 8-4 shows the output of the tag distribution.

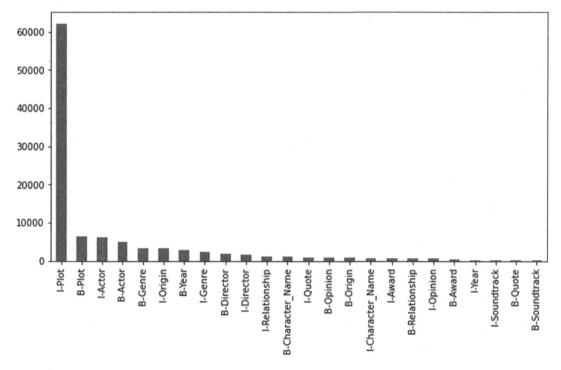

Figure 8-4. *Tag distribution*

Now, let's create a copy of train and test data for the further analysis and model build.

```
data=trivia_train.copy()
data1=trivia_test.copy()
```

Let's rename the columns using the following code.

```
data.rename(columns={"Sentence":"sentence_id","tokens":"words","tag":
"labels"}, inplace =True)
data1.rename(columns={"Sentence":"sentence_id","tokens":"words","tag":
"labels"}, inplace =True)
```

Model Building

Conditional Random Fields (CRF)

CRF is a conditional model class best suited to prediction tasks where the state of neighbors or contextual information affects the current prediction.

The main applications of CRFs are named-entity recognition, part of speech tagging, gene prediction, noise reduction, and object detection problems, to name a few.

In a sequence classification problem, the final goal is to find the probability of y (target) given input of sequence vector x.

Because conditional random fields are conditional models, they apply logistic regression to sequence data.

Conditional distribution is basically

$$Y = \text{argmax } P(y|x)$$

This finds the best output (probability) given sequence x.

In CRF, the input data is expected to be sequential, so we have to label the data as position i for the data point we are predicting.

We define feature functions for each variable; in this case, there is no POS tag. We only use one feature function.

The feature function's main purpose is to express a characteristic of the sequence that the data point represents.

Each feature function is relatively based on the label of the previous word and the current word. It is either a 0 or a 1.

So, to build CRF as we do in other models, assign some weights to each feature function and let weights update by optimization algorithm like gradient descent.

Maximum likelihood estimation is used to estimate parameters, where we take the negative log of distribution to make the derivative easier to calculate.

In general,

1. Define feature function.

2. Initialize weights to random values.

3. Apply gradient descent to parameter values to converge.

CRFs are more likely similar to logistic regression since they use *conditional probability distribution*. But, we extend the algorithm by applying feature functions as the sequential input.

We aim to extract entities from the given sentence and identifying their types using the CRF model. Now, let's import the required libraries.

We used the following libraries.

```
#For visualization
import matplotlib.pyplot as plt
```

```
import seaborn as sns
sns.set(color_codes=True)
sns.set(font_scale=1)
%matplotlib inline
%config InlineBackend.figure_format = 'svg'

# For Modeling
from sklearn.ensemble import RandomForestClassifier
from sklearn_crfsuite import CRF, scorers, metrics
from sklearn_crfsuite.metrics import flat_classification_report
from sklearn.metrics import classification_report, make_scorer

import scipy.stats
import eli5
```

Let's create grouped words and their corresponding tags as a tuple. Also, let's store the words of the same sentence in one list using the following sentence generator function.

```
class Get_Sent(object):

    def __init__(self, dataset):
        self.n_sent = 1
        self.dataset = dataset
        self.empty = False
        agg_func = lambda s: [(a, b) for a,b in zip(s["words"].values.tolist(),
                                            s["labels"].values.tolist())]
        self.grouped = self.dataset.groupby("sentence_id").apply(agg_func)
        self.sentences = [x for x in self.grouped]

    def get_next(self):
        try:
            s = self.grouped["Sentence: {}".format(self.n_sent)]
            self.n_sent += 1
            return s
        except:
            return None
```

```
# calling the Get_Sent function and passing the train dataset
Sent_get= Get_Sent(data)
sentences = Sent_get.sentences
```

So, the sentence looks like this.

Figure 8-5 shows the output of the Get_Sent function for training data.

```
[('steve', 'B-Actor'),
 ('mcqueen', 'I-Actor'),
 ('provided', 'O'),
 ('a', 'O'),
 ('thrilling', 'B-Plot'),
 ('motorcycle', 'I-Plot'),
 ('chase', 'I-Plot'),
 ('in', 'I-Plot'),
 ('this', 'I-Plot'),
 ('greatest', 'B-Opinion'),
 ('of', 'I-Opinion'),
 ('all', 'I-Opinion'),
 ('ww', 'B-Plot'),
 ('2', 'I-Plot'),
 ('prison', 'I-Plot'),
 ('escape', 'I-Plot'),
 ('movies', 'I-Plot')]
```

Figure 8-5. *Output*

```
# calling the Get_Sent function and passing the test dataset

Sent_get= Get_Sent(data1)
sentences1 = Sent_get.sentences

#This is what a sentence will look like.
print(sentences1[0])
```

Figure 8-6 shows the output of the Get_Sent function for test data.

```
[('i', 'O'), ('need', 'O'), ('that', 'O'), ('movie', 'O'), ('which', 'O'),
```

Figure 8-6. *Output*

For converting text into numeric arrays, we use simple and complex features.

Simple Feature Mapping

In this mapping, we have considered simple mapping of words and considered only six features of each word.

- word title
- word lower string
- word upper string
- length of word
- word numeric
- word alphabet

Let's look at the code for simple feature mapping.

```
# feature mapping for the classifier.

def create_ft(txt):
    return np.array([txt.istitle(), txt.islower(), txt.isupper(),
len(txt),txt.isdigit(),  txt.isalpha()])
```

```
#using the above function created to get the mapping of words for train
data.
```

```
words = [create_ft(x) for x in data["words"].values.tolist()]
```

```
#lets take unique labels
target = data["labels"].values.tolist()
```

```
#print few words with array
print(words[:5])
```

```
Output:
we got mapping of words as below (for first five words)
```

```
[array([0, 1, 0, 5, 0, 1]), array([0, 1, 0, 7, 0, 1]), array([0, 1, 0, 8,
0, 1]), array([0, 1, 0, 1, 0, 1]), array([0, 1, 0, 9, 0, 1])]
```

Likewise, we use the function created for the test data.

```
#using the above function created to get the mapping of words for test data.
words1 = [create_ft(x) for x in data1["words"].values.tolist()]
target1 = data1["labels"].values.tolist()
```

Apply five-fold cross validation for the random classifier model and get the results as follows. Next, the cross_val_predict function is used. It is defined in sklearn.

```
#importing package
from sklearn.model_selection import cross_val_predict
```

```
# train the RF model
Ner_prediction = cross_val_predict(RandomForestClassifier(n_
estimators=20),X=words, y=target, cv=10)
```

```
#import library
from sklearn.metrics import classification_report
```

```
#generate report
Accuracy_rpt= classification_report(y_pred= Ner_prediction, y_true=target)
print(Accuracy_rpt)
```

Figure 8-7 shows the classification report.

	precision	recall	f1-score	support
B-Actor	0.00	0.00	0.00	5010
B-Award	0.00	0.00	0.00	309
B-Character_Name	0.00	0.00	0.00	1024
B-Director	0.00	0.00	0.00	1787
B-Genre	0.00	0.00	0.00	3384
B-Opinion	0.00	0.00	0.00	810
B-Origin	0.00	0.00	0.00	779
B-Plot	0.00	0.00	0.00	6468
B-Quote	0.00	0.00	0.00	126
B-Relationship	0.00	0.00	0.00	580
B-Soundtrack	0.00	0.00	0.00	50
B-Year	0.88	0.99	0.93	2702
I-Actor	0.51	0.01	0.01	6121
I-Award	0.00	0.00	0.00	719
I-Character_Name	0.00	0.00	0.00	760
I-Director	0.00	0.00	0.00	1653
I-Genre	0.00	0.00	0.00	2283
I-Opinion	0.00	0.00	0.00	539
I-Origin	0.00	0.00	0.00	3340
I-Plot	0.45	0.47	0.46	62107
I-Quote	0.00	0.00	0.00	817
I-Relationship	0.00	0.00	0.00	1206
I-Soundtrack	0.00	0.00	0.00	158
I-Year	0.00	0.00	0.00	195
O	0.43	0.70	0.53	55895
accuracy			0.45	158822
macro avg	0.09	0.09	0.08	158822
weighted avg	0.36	0.45	0.38	158822

Figure 8-7. *Classification report*

Accuracy is 45%, and the overall F1 score is 0.38. The performance is not so good. Now let's further improve the accuracy, leveraging more features, like "pre" and "post" words.

Feature Mapping by Adding More Features

In this case, we consider the features of the next word and previous words corresponding to the given word.

1. Add a tag at the beginning of the sentence if the word starts the sentence.

2. Add a tag at the end of the sentence if the word ends the sentence. Also, consider the characteristics of the previous word and the next word.

Let's import the code for feature mapping.

First, upload the Python file to Colab or any working Python environment, as shown in Figure 8-8.

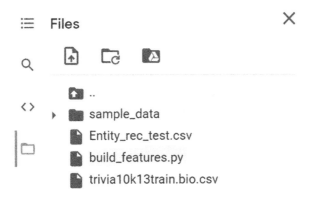

Figure 8-8. *Uploading Python script*

```
from build_features import text_to_features

# function to create features
def text2num(wrd):
    return [text_to_features(wrd, i) for i in range(len(wrd))]

# function to create labels
def text2lbl(wrd):
    return [label for token,label in wrd]
```

Let's prepare the training data features using sent2features function which in turn uses text to features function.

```
X = [text2num(x) for x in sentences]
```

Prepare the training data labels using `text2lbl` because the sentences are in a tuple consisting of words and tags.

```
y = [text2lbl(x) for x in sentences]
```

Similarly, prepare the test data.

```
test_X = [text2num(x) for x in sentences]
```

```
test_y = [text2lbl(x) for x in sentences]
```

Now that the training and test data are ready, let's initialize the model. First, let's initialize and build a CRF model without hyperparameter tuning.

```
#building the CRF model
ner_crf_model = CRF(algorithm='lbfgs',max_iterations=25)
```

```
#tgraining the model with cross validation of 10
ner_predictions = cross_val_predict(estimator= ner_crf_model, X=X, y=y, cv=10)
```

Let's evaluate the model on train data.

```
Accu_rpt = flat_classification_report(y_pred=ner_predictions, y_true=y)
print(Accu_rpt)
```

Figure 8-9 shows the classification report for train data.

	precision	recall	f1-score	support
B-Actor	0.94	0.92	0.93	5010
B-Award	0.71	0.63	0.67	309
B-Character_Name	0.72	0.39	0.51	1024
B-Director	0.85	0.81	0.83	1787
B-Genre	0.84	0.80	0.82	3384
B-Opinion	0.48	0.38	0.42	810
B-Origin	0.52	0.42	0.47	779
B-Plot	0.49	0.47	0.48	6468
B-Quote	0.78	0.37	0.51	126
B-Relationship	0.68	0.51	0.58	580
B-Soundtrack	0.69	0.22	0.33	50
B-Year	0.96	0.97	0.97	2702
I-Actor	0.94	0.92	0.93	6121
I-Award	0.81	0.73	0.77	719
I-Character_Name	0.71	0.40	0.51	760
I-Director	0.89	0.81	0.85	1653
I-Genre	0.77	0.72	0.74	2283
I-Opinion	0.26	0.12	0.16	539
I-Origin	0.69	0.68	0.68	3340
I-Plot	0.86	0.94	0.90	62107
I-Quote	0.78	0.44	0.56	817
I-Relationship	0.53	0.41	0.46	1206
I-Soundtrack	0.81	0.30	0.44	158
I-Year	0.62	0.66	0.64	195
O	0.87	0.84	0.85	55895
accuracy			0.84	158822
macro avg	0.73	0.59	0.64	158822
weighted avg	0.84	0.84	0.84	158822

Figure 8-9. *Classification report for train data*

Train data results: the overall F1 score is 0.84, and accuracy is 0.84.

Let's evaluate the model on test data.

```
#building the CRF model
crf_ner = CRF(algorithm='lbfgs',max_iterations=25)

#Fitting model on train data.
crf_ner.fit(X,y)
```

```
# prediction on test data
test_prediction=crf_ner.predict(test_X)
# get labels
lbs=list(crf_ner.classes_)
```

```
#get accuracy
metrics.flat_f1_score(test_y,test_prediction,average='weighted',labels=lbs)
```

Ouptut:

0.8369950502328535

```
#sort the labels
sorted_lbs=sorted(lbs,key= lambda name:(name[1:],name[0]))
```

```
#get classification report
print(metrics.flat_classification_report(test_y,test_prediction,
labels=sorted_lbs,digits=4))
```

Figure 8-10 shows the classification report for the test data.

	precision	recall	f1-score	support
O	0.8714	0.8423	0.8566	14143
B-Actor	0.9265	0.9207	0.9236	1274
I-Actor	0.9336	0.9227	0.9281	1553
B-Award	0.6923	0.6818	0.6870	66
I-Award	0.7941	0.7347	0.7633	147
B-Character_Name	0.7283	0.4452	0.5526	283
I-Character_Name	0.7453	0.5286	0.6186	227
B-Director	0.8575	0.8494	0.8534	425
I-Director	0.9116	0.8783	0.8947	411
B-Genre	0.8355	0.8048	0.8199	789
I-Genre	0.7853	0.7463	0.7653	544
B-Opinion	0.4867	0.3744	0.4232	195
I-Opinion	0.3333	0.1608	0.2170	143
B-Origin	0.4803	0.3842	0.4269	190
I-Origin	0.7088	0.6386	0.6719	808
B-Plot	0.4859	0.4705	0.4781	1577
I-Plot	0.8517	0.9318	0.8899	14661
B-Quote	0.8947	0.3617	0.5152	47
I-Quote	0.8315	0.4384	0.5741	349
B-Relationship	0.7812	0.5848	0.6689	171
I-Relationship	0.5665	0.4567	0.5057	289
B-Soundtrack	0.0000	0.0000	0.0000	8
I-Soundtrack	0.0000	0.0000	0.0000	30
B-Year	0.9683	0.9713	0.9698	661
I-Year	0.6098	0.5682	0.5882	44
accuracy			0.8412	39035
macro avg	0.6832	0.5879	0.6237	39035
weighted avg	0.8370	0.8412	0.8370	39035

Figure 8-10. *Classification report for test data*

The test data result shows an overall F1 score of 0.8370 and an accuracy of 0.8412. The accuracy increased by nearly 40% by adding more features, and the F1 score increased by 0.5. We can further increase the accuracy by performing hyperparameter tuning.

Let's take a random sentence and predict the tag using the trained model.

1. Create feature maps, as we did for training data, for input sentences using the word2feature function.

2. Convert into an array and predict using crf.predict(input_vector).

Next, convert each word into features.

```
def text_2_ftr_sample(words):
    return [text_to_features(words, i) for i in range(len(words))]
```

Split the sentences and convert each word into features.

```
#define sample sentence
X_sample=['alien invasion is the movie directed by christoper nollen'.
split()]

#convert to features
X_sample1=[text_2_ftr_sample(x) for x in X_sample]

#predicting the class
crf_ner.predict(X_sample1)
```

The following is the output.

```
[['B-Actor', 'I-Actor', 'O', 'O', 'O', 'O', 'O', 'B-Director',
'I-Director']]
```

Now, let's try the BERT model and check if it performs better than the CRF model.

BERT Transformer

BERT (Bidirectional Encoder Representations from Transformers) is a model that is trained on large data sets. This pretrained model can be fine-tuned as per the requirement and used for different tasks such as sentiment analysis, question answering system, sentence classification, and others. BERT transfers learning in NLP, and it is a state-of-the-art method.

BERT uses transformers, mainly the encoder part. The attention mechanism learns the contextual relationship between words and subwords. Unlike other models, Transformer's encoder learns all sequences at once. The input is a sequence of words (tokens) that are embed into vectors and then proceed to the neural networks. The output is the sequence of tokens that corresponds to the input token of the given sequence.

Let's implement the BERT model.

Import the required libraries.

```
#importing necessary libraries
from sklearn.preprocessing import LabelEncoder
from sklearn.model_selection import train_test_split
from sklearn.metrics import accuracy_score
from sklearn.metrics import classification_report, make_scorer

#importing NER models from simple transformers
from simpletransformers.ner import NERModel,NERArgs

#importing libraries for evaluation
from sklearn_crfsuite.metrics import flat_classification_report
from sklearn_crfsuite import CRF, scorers, metrics
```

Let's encode the sentence column using LabelEncoder.

```
#encoding sentence values
data["sentence_id"] = LabelEncoder().fit_transform(data["sentence_id"] )
data1["sentence_id"] = LabelEncoder().fit_transform(data1["sentence_id "] )
```

Let's convert all labels into uppercase.

```
#converting labels to upper string as it is required format
data["labels"] = data["labels"].str.upper()
data1["labels"] = data1["labels"].str.upper()
```

Next, separate the train and test data.

```
X= data[["sentence_id","words"]]
Y =data["labels"]
```

Then, create a train and test data frame.

```
#building up train and test data to dataframe
ner_tr_dt = pd.DataFrame({"sentence_id":data["sentence_id"],"words":data["w
ords"],"labels":data["labels"]})
test_data = pd.DataFrame({"sentence_id":data1["sentence_id"],"words":data1[
"words"],"labels":data1["labels"]})
```

Also, let's store the list of unique labels.

```
#label values
label = ner_tr_dt["labels"].unique().tolist()
```

We need to fine-tune the BERT model so that we can use parameters. Here, we changed epochs number and batch size. To improve the model further, we can change other parameters as well.

```
#fine tuning our model on custom data
args = NERArgs()
```

```
#set the # of epoch
args.num_train_epochs = 2
```

```
#learning reate
args.learning_rate = 1e-6
args.overwrite_output_dir =True
```

```
#train and evaluation batch size
args.train_batch_size = 6
args.eval_batch_size = 6
```

We now initialize the BERT model.

```
#initializing the model
Ner_bert_mdl= NERModel('bert', 'bert-base-cased',labels=label,args =args)
```

```
#training our model
Ner_bert_mdl.train_model(ner_tr_dt,eval_data = test_data,acc=accuracy_
score)
```

The eval_data is the data where loss is calculated. Figure 8-11 shows the output of training the BERT model.

The following is the output.

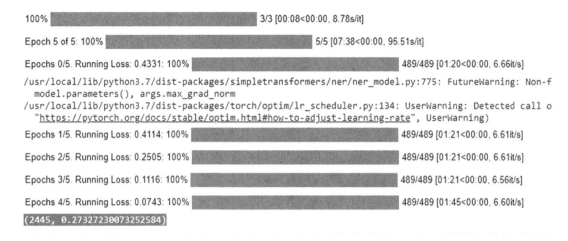

Figure 8-11. *Training a BERT model*

You can observe a loss of 0.27 in the final epoch.

```
#function to store labels and words of each sentence in list

class sent_generate(object):

    def __init__(self, data):
        self.n_sent = 1.0
        self.data = data
        self.empty = False
        fn_group = lambda s: [(a, b) for a,b in zip(s["words"].values.
        tolist(),s["labels"].values.tolist())]
        self.grouped = self.data.groupby("sentence_id").apply(fn_group)
        self.sentences = [x for x in self.grouped]

#storing words and labels of each sentence in single list of train data

Sent_get= sent_generate(ner_tr_dt)
sentences = Sent_get.sentences

#This is how a sentence will look like.
print(sentences[0])
def txt_2_lbs(sent):
    return [label for token,label in sent]

y_train_group = [txt_2_lbs(x) for x in sentences]
```

Figure 8-12 shows the output for sent_generate function for train data.

```
[('steve', 'B-ACTOR'), ('mcqueen', 'I-ACTOR'), ('provided', 'O'), ('a', 'O'), ('thrilling', 'B-PLOT'),
```

Figure 8-12. *Output*

```
#storing words and labels of each sentence in single list of test data
Sent_get= sent_generate(test_data)
sentences = Sent_get.sentences

#This is how a sentence will look like.
print(sentences[0])

def txt_2_lbs(sent):
    return [label for token,label in sent]

y_test = [txt_2_lbs(x) for x in sentences]
```

Figure 8-13 shows the output for the sent_generate function for the training data.

```
[('i', 'O'), ('need', 'O'), ('that', 'O'), ('movie', 'O'), ('which', 'O'),
```

Figure 8-13. *Output*

Let's evaluate the model on test data.

```
#evaluating on test data
result, model_outputs, preds_list = Ner_bert_mdl.eval_model(test_data)

#individual group report
accu_rpt = flat_classification_report(y_pred=preds_list, y_true=y_test)
print(accu_rpt)
```

Figure 8-14 shows the classification report for train data.

	precision	recall	f1-score	support
B-ACTOR	0.95	0.97	0.96	1274
B-AWARD	0.68	0.70	0.69	66
B-CHARACTER_NAME	0.71	0.72	0.72	283
B-DIRECTOR	0.88	0.92	0.90	425
B-GENRE	0.83	0.85	0.84	789
B-OPINION	0.47	0.52	0.50	195
B-ORIGIN	0.48	0.45	0.47	190
B-PLOT	0.54	0.52	0.53	1577
B-QUOTE	0.86	0.81	0.84	47
B-RELATIONSHIP	0.72	0.67	0.69	171
B-SOUNDTRACK	0.44	0.50	0.47	8
B-YEAR	0.98	0.98	0.98	661
I-ACTOR	0.96	0.96	0.96	1553
I-AWARD	0.77	0.81	0.79	147
I-CHARACTER_NAME	0.69	0.68	0.69	227
I-DIRECTOR	0.95	0.93	0.94	411
I-GENRE	0.81	0.76	0.79	544
I-OPINION	0.38	0.20	0.26	143
I-ORIGIN	0.70	0.75	0.72	808
I-PLOT	0.92	0.94	0.93	14661
I-QUOTE	0.91	0.79	0.85	349
I-RELATIONSHIP	0.63	0.57	0.60	289
I-SOUNDTRACK	0.48	0.37	0.42	30
I-YEAR	0.72	0.82	0.77	44
O	0.90	0.88	0.89	14143
accuracy			0.88	39035
macro avg	0.73	0.72	0.73	39035
weighted avg	0.88	0.88	0.88	39035

Figure 8-14. *Classification report*

The accuracy of test data is 88%. Let's evaluate the model on train data also to see if there is any overfitting.

```
#evaluating on train data
result_train, model_outputs_train, preds_list_train = Ner_bert_mdl.eval_
model(ner_tr_dt)
```

```
#individual group report of train data
```

```
report_train = flat_classification_report(y_pred=preds_list, y_true=y_
train_group)
print(report_train)
```

Figure 8-15 shows the classification report for train data.

	precision	recall	f1-score	support
B-ACTOR	1.00	1.00	1.00	5010
B-AWARD	0.97	0.97	0.97	309
B-CHARACTER_NAME	0.99	0.99	0.99	1024
B-DIRECTOR	0.94	0.99	0.97	1787
B-GENRE	0.97	0.95	0.96	3384
B-OPINION	0.83	0.94	0.88	810
B-ORIGIN	0.88	0.86	0.87	779
B-PLOT	0.94	0.93	0.94	6468
B-QUOTE	0.99	0.95	0.97	126
B-RELATIONSHIP	0.91	0.90	0.90	580
B-SOUNDTRACK	0.92	0.92	0.92	50
B-YEAR	1.00	1.00	1.00	2702
I-ACTOR	1.00	1.00	1.00	6121
I-AWARD	0.96	0.99	0.97	719
I-CHARACTER_NAME	0.99	0.99	0.99	760
I-DIRECTOR	1.00	1.00	1.00	1653
I-GENRE	0.96	0.90	0.93	2283
I-OPINION	0.98	0.98	0.98	539
I-ORIGIN	0.92	0.98	0.95	3340
I-PLOT	1.00	0.99	1.00	62107
I-QUOTE	0.99	1.00	0.99	817
I-RELATIONSHIP	0.94	0.97	0.95	1206
I-SOUNDTRACK	0.97	0.98	0.98	158
I-YEAR	0.88	0.97	0.92	195
O	0.99	0.99	0.99	55895
accuracy			0.98	158822
macro avg	0.96	0.97	0.96	158822
weighted avg	0.99	0.98	0.98	158822

Figure 8-15. *Classification report*

The accuracy of the training data is 98%. Compared to the CRF model, the BERT model is performing great.

Now, let's take a random sentence and predict the tag using the BERT model built.

```
prediction, model_output = Ner_bert_mdl.predict(["aliens invading is movie
by christoper nollen"])
```

Here, the model generates predictions that tag given words from each sentence, and model_output are outputs generated by the model.

Figure 8-16 shows the predictions with the random sentence.

```
[[{'aliens': 'B-PLOT'},
  {'invading': 'I-PLOT'},
  {'is': 'O'},
  {'movie': 'O'},
  {'by': 'O'},
  {'christoper': 'B-DIRECTOR'},
  {'nollen': 'I-DIRECTOR'}]]
```

Figure 8-16. *Model prediction*

Next Steps

- Add more features to the model example combination of words with gives a proper meaning for CRF model.

- The current implementation considers only two hyperparameters in the CV search for the CRF model; however, the CRF model offers more parameters that can be further tuned to improve the performance.

- Use an LSTM neural network and creating an ensemble model with CRF.

- For the BERT model, we can add a CRF layer to get a more efficient network as BERT finds efficient patterns between words.

Summary

We started this project by briefly describing NER and existing libraries to extract topics. Given the limitations, you have to build custom NER models. After defining the problem, we extracted movie names and director names and built an approach to solve the problem.

We tried three models: random forest, CRF, and BERT. We created the features from the text and passed them to RandomForest and CRF models with different features. As you saw, the more features better accuracy.

Instead of building these features manually, we used deep learning algorithms. We took a BERT-based pretrained model and trained NER on top of it. We did not use any manually created features, which is one of the advantages of deep learning models. Another positive point is pretrained models. You know how powerful transfer learning is, and the results of this project also showcase the same thing. We were able to produce good results from NER BERT when compared to handwritten features with CRF.

CHAPTER 9

Building a Chatbot Using Transfer Learning

In today's world, most businesses need to have customer support for their products and services. With the increase in e-commerce, telecommunication services, Internet-related products, and so on, the demand for customer service is only increasing. The nature of customer service support queries is repetitive in most conversations. Customer support conversations can be automated.

The following are some industries in which customer care is required.

- E-commerce

- Telecom

- Health care

- Manufacturing

- Electronic devices

The list is only going to increase as domains grow. So, what's the solution for this? Figure 9-1 shows the customer care application of chatbot.

Figure 9-1. *Applications of chatbots*

© Akshay Kulkarni, Adarsha Shivananda and Anoosh Kulkarni 2022
A. Kulkarni et al., *Natural Language Processing Projects*, https://doi.org/10.1007/978-1-4842-7386-9_9

Any business that focuses on the next-level customer experience needs to have outstanding after-sales service. Providing a reliable customer service call center to satisfy customers' questions and problems worked until a few years back. With the rise of the latest technology, customers are looking for greater convenience and speed in today's modern world. With restricted human resources, speed and the customer experience are the greatest challenges.

Most of these problems can be solved through chatbots if properly implemented. So, what is a chatbot?

Wikipedia defines a chatbot (also known as a spy, conversational bot, chatterbot, interactive agent, conversational interface, conversational AI, talkbot, or artificial spy entity) as "a computer program or an artificial intelligence which conducts a conversation via auditory or textual methods."

A *bot* is a computer's ability to understand human speech or text. A *chatbot* is a computer program that fundamentally simulates human conversations. It is short for *chat robot*.

Chatbots save time and effort by automating customer support. They are also used in other business tasks, such as collecting user information and organizing meetings. They can do a lot of things nowadays to make life a lot smoother.

Here are a few statistics.

- In 2020, more than 85% of customer interactions were managed without anyone.

- Companies with AI systems to help with customer support already see positive ROI, greater employee and customer satisfaction, and quicker time to resolution.

So, in this chapter, let's explore multiple ways to build a chatbot and look at how they can be implemented to solve business problems. There are so many ways we can build chatbots based on problem statements and data. If not everything, let's build a couple of them.

Approach

There are so many ways to build a chatbot. It depends on the complexity of the problem it should serve and the available data. Based on these considerations, there are the following types of chatbots.

Rule-based Chatbots

These are predefined keywords that the bot can understand. These commands must be written explicitly while coding using regular expressions or using any other text analysis method. It is very much rule-based, and if the user asks anything out of the box, the response from the bot is static, which shows that the bot cannot understand that input.

Even though it is very simple, it serves most of the problems where the task is repetitive, such as canceling the order or asking for a refund.

Generative or Smart Chatbots

These are advanced context-based chatbots that are backed by deep learning. There are no predefined sentences, but you should be able to answer most of the questions, if not all. At this stage, we cannot build a perfect chatbot, given the challenges in this area. But this is an active research area, and we can observe better results as days pass by.

We can also classify these chatbots into two more types based on their usage.

- **Vertical chatbots** are domain-specific chatbots focused on the application and do not work well enough across industries. For example, we are building a chatbot for doctors to answer questions about products on the market. However, we cannot use this for telecom industry applications.

- A **horizontal chatbot** is a general and open domain bot. The best examples are Siri, Alexa, and Google Assistant. These chatbots work at a high level and cannot be used for a granular task at the domain level. It does most of the tasks and acts as the starting point for most of the vertical bots.

Given these types of classification within chatbots, there are a lot of ways to implement these bots. There are a good number of frameworks to build chatbots for both vertical and horizontal chatbots. But let's go one level deeper and learn to implement these chatbots from scratch using natural language processing.

Also, let's not get into rule-based chatbots, given they are easy to implement. Let's investigate the different variations of advanced chatbots. We can use any of these chatbots based on the business problem.

So, let's start.

Chatbot 1: QA System Using Similarity Scores

One of the ways we can build chatbots is using the similarity scores between the sentences. We know that the bot always gets a sentence from the user. We need to use the sentence, and it should be answered so that the answer is relevant.

It must find a similar sentence in the data and show the response. Again, there are multiple ways to do this based on feature engineering techniques and formulas to calculate the similarity score. Figure 9-2 shows the architecture for this simple chatbot.

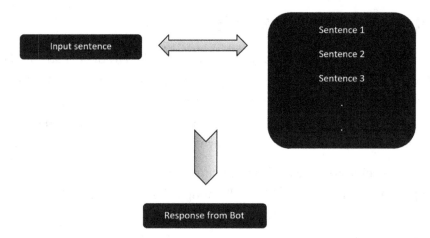

Figure 9-2. *Simple chatbot flow*

To convert text to features, we can use either a count vectorizer or TF-IDF vectorizer. In this exercise, let's use TF-IDF since we know it's going to perform better. We can also use word embeddings.

So, let's implement this.

```
#import packages
import pandas as pd
import numpy as np

#import nltk
import nltk

#import other
import string
import random
```

```
#importing the data

data=open('data.txt','r',errors = 'ignore')
text=data.read()
print(text)
```

Figure 9-3 shows the text imported from Wikipedia, and it has all the sentences in it. Since we are doing this at the sentence level, let's convert them to sentences. And convert them to words to obtain the features using TF-IDF.

```
Data science is an interdisciplinary field that uses scientific methods, processes, algorithms and systems to extract knowledge
and insights from structured and unstructured data,[1][2] and apply knowledge and actionable insights from data across a broad
range of application domains. Data science is related to data mining, machine learning and big data.

Data science is a "concept to unify statistics, data analysis, informatics, and their related methods" in order to "understand
and analyze actual phenomena" with data.[3] It uses techniques and theories drawn from many fields within the context of mathem
atics, statistics, computer science, information science, and domain knowledge. However, data science is different from compute
r science and information science. Turing Award winner Jim Gray imagined data science as a "fourth paradigm" of science (empiri
cal, theoretical, computational, and now data-driven) and asserted that "everything about science is changing because of the im
pact of information technology" and the data deluge.[4][5]

Contents
1        Foundations
1.1      Relationship to statistics
2        Etymology
2.1      Early usage
2.2      Modern usage
3        Impact
4        Technologies and techniques
4.1      Techniques
5        See also
6        References
```

Figure 9-3. *Text from Wikipedia*

```
# to lower  case
text=text.lower()

# tokenize the words
wr_ids = nltk.word_tokenize(text)

# tokenize the sent
st_ids = nltk.sent_tokenize(text)

wr_ids[0]
'data'

st_ids[0]
```

The following is the output.

```
'data science is an interdisciplinary field that uses scientific methods,
processes, algorithms and systems to extract knowledge and insights from
structured and unstructured data,[1][2] and apply knowledge and actionable
insights from data across a broad range of application domains.'
```

We need to calculate the similarity between sentences, and for that, distance measures are necessary. The most used distance metric is cosine similarity. Let's import the libraries for that and use them to calculate distance.

```
#import libraries for distance calculation
from sklearn.feature_extraction.text import TfidfVectorizer
from sklearn.metrics.pairwise import cosine_similarity
```

The following code is the response function, which is called once there is input from the bot. First, the code takes the input and appends it to the text corpus. Then TF-IDF is calculated for all the tokens using TfidfVectorizer. Next, the similarity score is calculated using cosine similarity between input and all the sentences present in the corpus. Then the sentence with the highest similarity is given a response from the bot.

If the TF-IDF is 0, then the standard response is given.

```
#define function for response

def get_output(user_input):

    #define the output
    output=''

    #append input to text
    st_ids.append(user_input)

    #define tfidf
    txt_v = TfidfVectorizer(stop_words='english')

    #get vector
    vec_txt = txt_v.fit_transform(st_ids)

    #get score
    rank_score= cosine_similarity(vec_txt[-1], vec_txt)
```

```
    idx=rank_score.argsort()[0][-2]
    ft_out = rank_score.flatten()
    ft_out.sort()
    final_v = ft_out[-2]

    if(final_v==0):
        output=output+"Dont know this annswer, Ask something else"
        return output
    else:
        output = output+st_ids[idx]
        return output
```

The final piece of code activates the bot, gets the input, calls the response function, and gives the output.

```
# Final code to run the bot

print("Enter your question")
print("")
in_txt = input()
in_txt=in_txt.lower()
print("")
print("Ans:", get_output(in_txt))
print("")
st_ids.remove(in_txt)
```

The following is the output.

Done. Figure 9-4 shows the model output for the question "What is data science?"

```
Enter your question

What is Data Science?

Ans: however, data science is different from computer science and information science.
```

Figure 9-4. *Model output*

We implemented a simple chatbot using TFIDF and similarity scores. Many things can be implemented on top of this, but there are the fundamentals behind these types of bots. Post this, APIs need to be created. The front end has to be integrated and hosted in a repository. All these things are the engineering or dev ops side of things.

The following are the disadvantages of these kinds of Q&A systems.

- The chatbot can answer the questions only on data present in the input text we used in the initial stage of the code. It fails to answer the question out of it.

- Fails to capture the entire context of the question since we did not use any advanced text to feature techniques.

But these are well suited for vertical chatbots where we know what the backed data is and certainly know that questions are within that. A chatbot can be built for legal activities using all possible legal documents as training data.

So that ends one type of chatbot implementation. It fails to understand the context. This is a problem given that the chatbot has to understand the context of the question; otherwise, users will not be excited about it.

This leads to our second chatbot built using deep learning, which captures the context.

Chatbot 2: Context-based Chatbot Using a Pretrained Model

Let's explore one more variation of chatbots. This is general-purpose and builds on a huge data set after training for a long time using GPUs. But let's say we don't have the resources to do this. Then comes the concept of transfer learning. Let's use one of the pretrained models and solve the problem at hand.

We already saw the capabilities of deep learning to capture context and increase accuracy. Let's look at another example of using a deep learning-based pretrained model to improve the chatbots. So, let's explore.

Hugging Face Transformers

Let's use a state-of-the-art Hugging Face library for this task. The *transformers* are an open source library that has extraordinary pretrained models which can be downloaded and build on any downstream applications—just like that. It's very easy to use, and the results are great.

So let's install and start using it.

```
#install transformers
!pip install transformers
```

Now comes model selection. We all know how effective the BERT architecture is to produce great contextual outputs. Let's consider one of the BERT-based models for this task.

Let's import the model and the tokenizer.

```
#import model and tokenizer
from transformers import BertForQuestionAnswering
from transformers import BertTokenizer

# import torch
import torch
```

Let's load the pretrained model. We can simply change these Q&A-based pretrained models to any other present on the Hugging Face website, and it works.

```
#loading the pre trained Model
qna_model = BertForQuestionAnswering.from_pretrained('bert-large-uncased-
whole-word-masking-finetuned-squad')

#loading the Tokenizer for same model
qna_tokenizer = BertTokenizer.from_pretrained('bert-large-uncased-whole-
word-masking-finetuned-squad')
```

To ask questions, we first need text. Let's take a paragraph from Wikipedia based on cricketer Virat Kohli. Figure 9-5 shows a snapshot of the text.

```
ans_text = """Virat Kohli (Hindustani: [ʋɪˈraːt̪ ˈkoːɦliː] (Abo
ut this soundlisten); born 5 November 1988) is an Indian crick
eter and the current captain of the India national team. A rig
ht-handed top-
order batsman. He is widely regarded as one of the greatest ba
tsman of present era. He plays for Delhi in domestic cricket a
nd for Royal Challengers Bangalore in the Indian Premier Leagu
e (IPL) as captain of the franchise since 2013. He was a part
of Indian cricket team which won the 2011 crickket world cup

Kohli captained India Under-19s to victory at the 2008 Under-
19 World Cup in Malaysia. After a few months later, he made hi
s ODI debut for India against Sri Lanka at the age of 19. Init
ially having played as a reserve batsman in the Indian team, h
e soon established himself as a regular in the ODI middle-
order and was part of the squad that won the 2011 World Cup. H
e made his Test debut in 2011 and shrugged off the tag of "ODI
 specialist" by 2013 with Test hundreds in Australia and South
 Africa.[3] Having reached the number one spot in the ICC rank
ings for ODI batsmen for the first time in 2013,[4] Kohli also
 found success in the Twenty20 format, winning the Man of the
Tournament twice at the ICC World Twenty20 (in 2014 and 2016).
"""
```

Figure 9-5. *Text from Wikipedia*

Now let's build a function that takes user input questions, hits the text, and predicts answers.

```
#function to get an answer for a use given question

def QnA(user_input_que):

  #tokinizng the texts
  in_tok = qna_tokenizer.encode_plus(user_input_que, ans_text, return_
  tensors="pt")

  #getting scores from tokens
  ans_str_sc, ans_en_sc = qna_model(**in_tok,return_dict=False)
```

```
#getting the position
ans_st = torch.argmax(ans_str_sc)
ans_en = torch.argmax(ans_en_sc) + 1

#ids are then converted to tokens
ans_tok = qna_tokenizer.convert_ids_to_tokens(in_tok["input_ids"][0]
[ans_st:ans_en])

#getting the answer
return qna_tokenizer.convert_tokens_to_string(ans_tok)
```

Let's take a couple of questions to see how it works.

```
Example 1:

que = "when did kohli win world cup"

QnA(que)
```

2011 world cup

```
Example 2:

que = "when did kohli born"

QnA(que)
```

5 november 1988

Both answers are spot on. That's how powerful these pretrained models which capture the context so well.

Now let's take another example from a different domain and see if it can perform to the same level. Let's pick a healthcare-related text. This text is also from Wikipedia. Figure 9-6 shows the snapshot for the text.

```
ans_text = """Coronavirus disease 2019 (COVID-
19) is a contagious disease caused by severe acute respiratory
 syndrome coronavirus 2 (SARS-CoV-
2). The first known case was identified in Wuhan, China, in De
cember 2019.[7] The disease has since spread worldwide, leadin
g to an ongoing pandemic.[8]

Symptoms of COVID-
19 are variable, but often include fever,[9] cough, headache,[
10] fatigue, breathing difficulties, and loss of smell and tas
te.[11][12][13] Symptoms may begin one to fourteen days after
exposure to the virus. At least a third of people who are infe
cted do not develop noticeable symptoms.[14] Of those people w
ho develop symptoms noticeable enough to be classed as patient
s, most (81%) develop mild to moderate symptoms (up to mild pn
eumonia), while 14% develop severe symptoms (dyspnea, hypoxia,
 or more than 50% lung involvement on imaging), and 5% suffer
critical symptoms (respiratory failure, shock, or multiorgan d
ysfunction).[15] Older people are at a higher risk of developi
ng severe symptoms. Some people continue to experience a range
 of effects (long COVID) for months after recovery, and damage
 to organs has been observed.[16] Multi-
year studies are underway to further investigate the long-
term effects of the disease.[16]

COVID-
19 transmits when people breathe in air contaminated by drople
ts and small airborne particles containing the virus. The risk
 of breathing these in is highest when people are in close pro
ximity, but they can be inhaled over longer distances, particu
larly indoors. Transmission can also occur if splashed or spra
yed with contaminated fluids in the eyes, nose or mouth, and,
rarely, via contaminated surfaces. People remain contagious fo
r up to 20 days, and can spread the virus even if they do not
develop symptoms.[17][18]"""
```

Figure 9-6. *Text from Wikipedia on covid19*

Let's ask questions now.

Question 1:

que = "where did covid started?"

QnA(que)

wuhan , china

Question 2:

que = "what are the symptoms of covid19?"

QnA(que)

fever , [9] cough , headache , [10] fatigue , breathing difficulties , and loss of smell and taste

Those are great answers, isn't it? We can check out different pretrained models and compare the results.

Now, let's move on to one more pretrained model based on RNN.

Chatbot 3: Pretrained Chatbot Using RNN

With less data, the response from the bot is not going to be that great because it struggles to understand the context. Also, training the bot with a lot of data is a challenge for each use case. That is where we use transfer learning.

Someone already trained the algorithm with a certain amount of data. That model is available for everyone to use. Generalized pretrained models solve many of the problems across industries.

One of those examples is a chatbot trained using a language model. The one way of solving this problem is generating a sequence of text or next word or character given all previous words and characters. These models are called *language models*.

Usually, recurrent neural networks (RNN) train a model because they are very powerful and expressive through remembering and processing past information through their high dimensional hidden state units.

There are two types of models.

- **Word level**: The words are used as input to train the model. If a particular word Is not present in the corpus, we do not get a prediction for that.

- **Character level**: The training is done at the character level.

In this architecture, there are certain pretrained models. Let's use them for our application.

Next, let's explore implementing RNN. The following is the GitHub link pretrained chatbot using character-level RNN. It's been trained using a lot of Reddit data.

Github link: `https://github.com/pender/chatbot-rnn`

Close this repository or download the whole project and keep it in your local. The following are the steps to use this pretrained model with RNN architecture.

1. Download the pretrained model.

   First, let's download the pretrained model, which is trained on Reddit data.

   Link: `https://drive.google.com/uc?id=1rRRY-y1KdVk4UB5qhu7 BjQHtfadIOmMk&export=download`

2. Unzip the downloaded pretrained model.

3. Place the unzipped folder into the "models" folder present in the downloaded project as shown in Figure 9-7.

data	04-02-2018 23:11	File folder	
models	20-04-2019 11:44	File folder	
reddit-parse	04-02-2018 23:11	File folder	
chatbot	04-02-2018 23:11	Python File	16 KB
LICENSE	04-02-2018 23:11	Text Document	2 KB
model	04-02-2018 23:11	Python File	15 KB
README.md	04-02-2018 23:11	MD File	10 KB
train	04-02-2018 23:11	Python File	11 KB
utils	04-02-2018 23:11	Python File	11 KB

Figure 9-7. *Repo structure*

4. Open the terminal and navigate to the folder where the project is saved locally.

5. Run the Python chatbot model, as shown in Figure 9-8.

    ```
    python chatbot.py
    ```

```
\chat bot\chatbot-rnn-master>python chatbot.py
Creating model...
WARNING:tensorflow:From C:\Users\adarsha.shivananda\AppData\Local\Continuum\anaconda3\lib\site-packages\ter
y.py:263: colocate_with (from tensorflow.python.framework.ops) is deprecated and will be removed in a futur
Instructions for updating:
Colocations handled automatically by placer.
WARNING:tensorflow:From C:\Users\adarsha.shivananda\Desktop\Pro\DS Book\Book4\Chapters\Completed Chapters\c
: GRUCell.__init__ (from tensorflow.python.ops.rnn_cell_impl) is deprecated and will be removed in a future
Instructions for updating:
This class is equivalent as tf.keras.layers.GRUCell, and will be replaced by that in Tensorflow 2.0.
WARNING:tensorflow:From C:\Users\adarsha.shivananda\Desktop\Pro\DS Book\Book4\Chapters\Completed Chapters\c
5: dynamic_rnn (from tensorflow.python.ops.rnn) is deprecated and will be removed in a future version.
Instructions for updating:
Please use `keras.layers.RNN(cell)`, which is equivalent to this API
Restoring weights...
WARNING:tensorflow:From C:\Users\adarsha.shivananda\AppData\Local\Continuum\anaconda3\lib\site-packages\ter
 checkpoint_exists (from tensorflow.python.training.checkpoint_management) is deprecated and will be remove
Instructions for updating:
Use standard file APIs to check for files with this prefix.
```

Figure 9-8. *Loading the model*

Let's ask a few questions to see how it works. Figure 9-9 shows the output.

```
> hey
  Hey! I'm not your brother!
> then who are you?
  No! Nothing!
> what do you do?
  Tell me your answer to this question.
> you chat
  What are you talking about?
```

Figure 9-9. *Model output*

Again, this is a generalized bot that is built on Reddit data. We need to retrain this on whatever domain we are using it too. We can use the train.py to train this model on our data set.

Information on the RNN that trained this model is at http://karpathy.github.io/2015/05/21/rnn-effectiveness/.

Future Scope

We tried out various models to build the chatbots in the back end. But chatbot also requires the frameworks to make the conversation smooth and logical. We can use the following frameworks for implementation.

RASA

RASA (`https://rasa.com`) is an AI framework to build chatbots. It's open source and low code framework and chatbots can be built with very little effort. It can do both understanding the text and create flows for chats.

RASA stack has a bunch of functionalities to build effective chatbots. It lets the user train the custom models where industry-specific data builds the chatbots. RASA already does the heavy lifting of building the framework.

RASA's architecture has two main components: RASA NLU to understand the incoming text where it identifies intents and RASA CORE which gives the output for the question asked by the user.

Microsoft Bot Framework

MS bot framework is another framework to create bots as well deploy them to azure services. It has all the end-to-end components to build the bots and connect to any platform like skype, SMS, and so forth.

This framework is open source. You can use the SDK at `https://github.com/microsoft/botframework-sdk` to create the chatbots.

The bot builder SDK has two other components: a developer portal and a directory that help you easily develop and deploy.

It creates a very effective dialogue flow. For example, one of the main objectives of the bots is to remember what happened in the previous action. It has to remember the historical flows to give contextual answers.

Conclusion

We explored multiple ways of building a chatbot in this chapter using the power of natural language processing and deep learning. This is the back end of the chatbot and integrating this with a front end is a task.

We built a simple chatbot using a similarity score where the sentences are converted to TF-IDF features and distance is calculated using cosine similarity. In another approach, a deep neural network was leveraged to build the context-based chatbots. Finally, you saw how RNN can predict the next word and how it works in chatbots using a pretrained model. There is active research in this field, and we can expect many more groundbreaking results that take customer experience to another level.

There is a huge market for chatbots, given that it has already proved its credibility. It's only going to increase in coming years with applications both vertically and horizontally making the customer's life a lot easier.

CHAPTER 10

News Headline Summarization

Text summarization is all about converting a bunch of paragraphs into a few sentences that explain the whole document's gist. There are hundreds of applications in every industry, given the amount of text data. Text data is increasing exponentially. A lot of time is needed to analyze, understand, and summarize each piece of it.

At some point in time, we may need a smart way to summarize text data, which is why it's a very active research area. We can achieve some level of success, but there is a long way to go since capturing context like a human being is easier said than done!

A summarization can be of two types based on size.

- Long summary: A paragraph with 10 sentences or more.

- Short summary: A sentence or a two with 10–20 words.

Summarization has abundant applications across industries. It can be leveraged in various places, including the following.

- Summarizing news articles to enrich the user experience

- Legal document summarization

- Summarizing clinical research documents

- Product review insight from multiple data sources

- Call center data summarization to understand the high-level cause

- Summarize the educational notes and book for quick revision

- Transcript summarization

- Social media analysis

© Akshay Kulkarni, Adarsha Shivananda and Anoosh Kulkarni 2022
A. Kulkarni et al., *Natural Language Processing Projects*, https://doi.org/10.1007/978-1-4842-7386-9_10

Let's look at the various ways of summarizing text and its challenges and implement a few state-of-the-art techniques in this chapter.

Approach

Let's first understand the different types of summarization techniques before getting into implementation.

There are two main ways to summarize text in NLP: *extractive summarization* and *abstractive summarization*.

Extractive Summarization

Extractive summarization is the simple technique that selects the important sentences from the document and creates the summary. In the extractive text summarization technique, the most repeating sentence gets more importance and generates a summary. The original text remains the same, and without making any changes to it, the summary would be generated. But the disadvantage is, the summary might not be as perfect as humans write.

Here is a simple example.

- Source text: Varun went to attend an event in Bangalore about data science. The event discussed how machine learning works along with the applications of AI and ML in the retail industry, including churn prediction, recommendation system, and buying propensity model. Finally, the session ended with next-generation AI.

- Extractive summary: Varun went to attend an event about data science. It extracted few important words from the source, and that's it.

There are different types of extractive summarization techniques based on the algorithm used.

- GraphBase, which uses the graph technology, creates nodes and edges using the document and finally finding relations to summarize it. TextRank and LexRank is an example of graph-based algorithms.

- Feature-based algorithms extract certain features from each sentence of the document and then decide the important sentence based on the features. The feature can be term frequency, presence of verb, and so forth. TextTeaser and Luhn's Algorithm are different feature-based algorithms.

- Topic-based algorithms extract topics from the documents using algorithms like singular value decomposition and score the sentences based on the topics extracted. Gensim has the lsimodel function to carry out topic-based summarization.

Implementing extractive summarization techniques is straightforward. For more information, please refer to our book *Natural Language Processing Recipes: Unlocking Text Data with Machine Learning and Deep Learning Using Python* (Apress, 2019).

Abstractive Summarization

Abstract techniques involve interpreting and shortening the parts of the source document. This algorithm creates new sentences and phrases which cover most of the context from the original text. It's more difficult to develop these kinds of algorithms, so extraction is still very popular.

Here is a simple example.

- Source text: Varun went to attend an event in Bangalore about data science. The event discussed how machine learning works along with the applications of AI and ML in the retail industry, like churn prediction, recommendation system, and buying propensity model. Finally, the session ended with next-generation AI.

- Abstractive summary: Varun attended a data science event where machine learning and its application was discussed, including how to predict customer churn in the retail industry and build recommendation system along with the future of AI.

So, if you look at it, the summary is handcrafted using the context of the source. It did not only extract a few important sentences, like in the extractive summary.

There have been many approaches to build abstractive summarization. And with the advancements in deep learning, we are now able to achieve good results.

This chapter looks at implementing an abstractive summarization technique using deep learning algorithms. You also learn how to leverage pretrained models to increase accuracy.

Figure 10-1 illustrates the approach design to solve a text summarization problem.

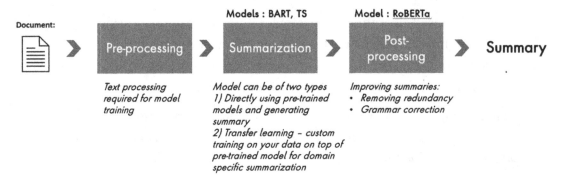

Figure 10-1. *Approach design*

More information on the models and process are uncovered as we move along.

Environment Setup

Since it takes a lot of resources to train the model, let's use Google Colaboratory (Colab) for this experience (see Figure 10-2).

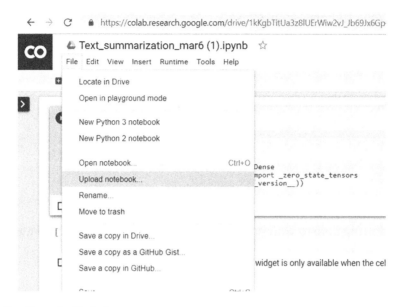

Figure 10-2. Google Colab

Setting up and using is as easy as it gets. Go to `https://colab.research.google.com/notebooks/welcome.ipynb` to start Colab. Then copy and paste the codes from or import the notebook.

Go to File and select **Upload notebook** to upload the notebooks downloaded from this chapter's Github link for this exercise.

Understanding the Data

We are using a news data set to generate headlines. The data set consists of text and headlines for more than 3000 records.

Let's start with importing the required packages.

```
#Import packages

#data processing
import pandas as pd
import numpy as np

#For text extraction
import re
from bs4 import BeautifulSoup
```

```
from nltk.corpus import stopwords
import time

pd.set_option("display.max_colwidth", 200)

#For modeling
#import tf
import tensorflow as tf

# import keras
from keras.preprocessing.text import Tokenizer
```

Before importing data to your current instance, we need to import data to Colab through a set of codes.

Follow these steps to import the data.

```
# Import data

from google.colab import files
uploaded = files.upload()
```

You get the pop-up shown in Figure 10-3. Choose the directory to upload the file for this exercise.

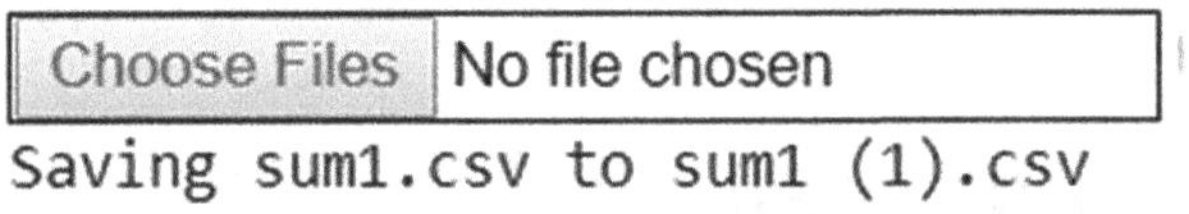

Choose Files	No file chosen

Saving sum1.csv to sum1 (1).csv

Figure 10-3. *Choose file pop-up*

```
#Import the file to the current session

#Importing data

df=pd.read_csv('sum1.csv')
df.head()
```

Figure 10-4 shows the df output.

	headlines	text
0	upGrad learner switches to career in ML & AI w...	Saurav Kant, an alumnus of upGrad and IIIT-B's...
1	Delhi techie wins free food from Swiggy for on...	Kunal Shah's credit card bill payment platform...
2	New Zealand end Rohit Sharma-led India's 12-ma...	New Zealand defeated India by 8 wickets in the...
3	Aegon life iTerm insurance plan helps customer...	With Aegon Life iTerm Insurance plan, customer...
4	Have known Hirani for yrs, what if MeToo claim...	Speaking about the sexual harassment allegatio...

Figure 10-4. *Output*

Note   Every time the session restarts, data has to be imported again.

```
# Rows and columns in the dataset
df.shape
```

```
(3684, 2)
```

```
# Check for any null values
df.isnull().sum()
```

```
headlines    0
text         0
dtype: int64
```

There are 3684 rows and two columns with no null values.

Text Preprocessing

There is a lot of noise present in the text data. Text preprocessing is vital. Let's clean it up.

Text preprocessing tasks include

- Converting the text to lower case

- Removing punctuations

- Removing numbers

- Removing extra whitespace

Let's create a function that includes all these steps so that we can use this wherever required in future instances (for model training)

```
#Text preprocessing function

def txt_preprocessing(txt):
    txt = re.sub(r'[_"\-;%()|+&=*%.,!?:#$@\[\]/]', ' ', txt)
    txt = re.sub(r'\'', ' ', txt)
    txt = txt.lower()
    return txt

headlines_processed = []
clean_news = []

# headline text cleaning
for z in df.headlines:
    headlines_processed.append(txt_preprocessing(z))

# news text cleaning
for z in df.text:
    clean_news.append(txt_preprocessing(z))
```

Now that most of the data understanding and preprocessing are completed let's move into the model building phase.

Model Building

Before we get into model building, let's discuss transfer learning.

The process where the model is trained to do a specific task and the knowledge gained in this model is leveraged to do another task on different data set is called *transfer learning*.

The language models require huge data and resources to give better accuracy, and every individual or company would not have access to both. The big companies like Google, Microsoft, and Facebook train these algorithms with their abundant data and research teams and make them open source. We can simply download and start utilizing those pretrained models for specific tasks.

But the challenge is, since these models are trained on a wide variety of data, the domain-specific tasks might not work that well. That's where we use domain-specific data and re-train or customize the models on top of existing ones.

This whole process is transfer learning.

We are using 2 LSTM based pretrained models in this project. In 2019, Google and Facebook published new transformer papers on T5 and BART. Both papers reported great performance in tasks like abstractive summarization. We use these two transformers to see what the results look like on the data set.

First, use a pretrained model directly to generate a summary for any input brief (a small demo). In this case, let's generate a news summary.

```
#connect to hugging face
!git clone https://github.com/huggingface/transformers \
&& cd transformers \

#install
!pip install -q ./transformers

#Import
import torch
import transformers
from transformers import BartTokenizer, BartForConditionalGeneration

torch_device = 'cpu'
tokenizer = BartTokenizer.from_pretrained("facebook/bart-large")
model = BartForConditionalGeneration.from_pretrained("facebook/bart-large")

#Function to summarize
def summarize_news(input_text, maximum_length, minimum_length):

  #Tokenize
  input_txt_ids = tokenizer.batch_encode_plus([input_text], return_
  tensors='pt', max_length=1024)['input_ids'].to(torch_device)

  #summarize
  ids_sum = model.generate(input_txt_ids, max_length=int(maximum_length),
  min_length=int(minimum_length))

  #get the text summary
  output_sum = tokenizer.decode(ids_sum.squeeze(), skip_special_
  tokens=True)
  return output_sum
```

```
input_text = "According to a Hindustan Times report, the 14 Keralites
were among the terrorists and militants freed by the Taliban from Bagram
jail. As of now, unconfirmed reports state that two Pakistani residents
were detained by the Sunni Pashtun terrorist group for trying to blow off
an Improvised Explosive Device (IED) device outside Turkmenistan embassy
in Kabul on August 26. And as intelligence reports indicate, an IED was
recovered from the two Pakistani nationals soon after the Kabul airport
blast. As per reports, a Kerala resident contacted his home, while the
remaining 13 are still in Kabul with the ISIS-K terrorist group. After
Syria and Levant occupied Mosul in 2014, people from the Malappuram,
Kasaragod and Kannur districts left India and joined the jihadist group in
West Asia from where a few Keralites came down to Nangarhar province of
Afghanistan."

#Generate summary using function
summarize_news(input_text,20,10)

#output
14 Keralites were among the terrorists and militants freed by the Taliban
from Bagram
```

This looks good; the pretrained BART model did provide a reasonable summary. However, if you want to further improve it and bring in domain-specific context for a better summary, you can use this pretrained model as a base and perform transfer learning (custom training on your dataset). Let's explore how to do that in the next section.

How do you train the summarization model? First, let's discuss a few more concepts. Broadly, sequence models can be of three types.

- seq2num

- seq2class

- seq2seq

Since in summarization, the input is in the form of sequence text (news), and even output is in the form of sequence text again (news headline). This is a seq2seq model, which is popular in applying text summarization, text generation, machine translation, and Q&A.

Now let's discuss the building blocks of seq2seq models.

The first thing that comes to mind when we hear about sequence is LSTM. The LSTM architecture captures sequence. So, keep it as a base. The following are some of the building blocks.

- **Bidirectional LSTM**: The first layer of the recurrent network is replicated to form another architecture where the reverse of the input is passed through it to capture the relationship of words and context in both directions, capturing better context.

- **Autoencoders**: Encoder and decoder architecture. The encoder encodes the input text and generates the fixed-length vector, which is fed into the decoder layer for prediction. The main disadvantage here is that the decoder has visibility to the fixed-length vector-only and might be limited to some context, and there is a high chance of important information loss. Hence the birth of the attention mechanism.

- **Attention mechanism**: The attention mechanism captures the importance each word carries to generate the desired outcome. It increases the importance of few words that result in the desired output summary or sequence.

There is also self-attention and transformers. Getting into the details of all these concepts is beyond the scope of this book. Please refer to the respective research papers to gain a deeper understanding of these concepts.

Now that you understand how seq2seq models are formulated, let's discuss how to build them. There are three options.

- Training from scratch (requires a large amount of data and resources to achieve better accuracy)

- Only using a pretrained model

- Transfer learning (custom training on a pretrained model)

Let's jump into BART, which is built on the building blocks we have discussed.

BART: Simple-Transformer Pretrained Model

Bidirectional encoders form the first part of the BART transformers architecture. Autoregressive decoders form the second part. They are integrated to devise the seq2seq model. And, hence BART can also be used for many text-to-text-based applications, including language translation and text generation.

The AR in BART stands for *autoregressive*, where the model consumes the previously generated output as an additional input at each step.

This makes BART particularly effective in text generation tasks. We fine-tuned BART for abstractive text summarization. BART offers an abstractive summarizer that can intelligently paraphrase, capture, and combine information based on contextual overlap. You can also experiment with tuning the beam width hyperparameter to optimize generative capabilities while minimizing the chances of misinformation.

BART pretrained model is trained on CNN/*Daily Mail* data for the summarization task.

Please note that you can also use this pretrained model directly without customization and generate the summary output.

Let's start the implementation using this model. First, install the necessary libraries.

```
#Installing transformers
!pip install simpletransformers
!pip install transformers

#Importing necessary libraries from simple transformers
import pandas as pd
from simpletransformers.seq2seq import Seq2SeqModel,Seq2SeqArgs
```

BART accepts two columns: target_text and input_text. Let's rename the existing columns to match BART's requirements.

```
#Rename columns as per pretrained model required format
df=df.rename(columns={'headlines':'target_text','text':'input_text'})

model_args = Seq2SeqArgs()

#Initializing number of epochs
model_args.num_train_epochs = 25

#Initializing no_save arg
model_args.no_save = True
```

```
#Initializing evaluation args
model_args.evaluate_generated_text = True
model_args.evaluate_during_training = True
model_args.evaluate_during_training_verbose = True
```

Let's use the Seq2SeqModel function to train the BART model on the training data. We are using the BART-large pretrained model here.

```
# Initialize the model with type as 'bart' and provide model args

model = Seq2SeqModel(
    encoder_decoder_type="bart",
    encoder_decoder_name="facebook/bart-large",
    args=model_args,
    use_cuda=False,
)
```

```
 #Splitting data into train-test

from sklearn.model_selection import train_test_split

train_df, test_df = train_test_split(df, test_size=0.2)
train_df.shape, test_df.shape
```

```
#Training the model and keeping eval dataset as test data

model.train_model(train_df, eval_data=test_df)
```

Figure 10-5 shows the model training.

100%	2946/2946 [00:06<00:00, 294.59it/s]
Epoch 10 of 10: 100%	10/10 [39:58<00:00, 238.45s/it]
Epochs 0/10. Running Loss: 1.6402: 100%	369/369 [03:10<00:00, 2.10it/s]
100%	737/737 [00:01<00:00, 376.83it/s]
Generating outputs: 100%	93/93 [00:30<00:00, 3.69it/s]
Epochs 1/10. Running Loss: 0.8489: 100%	369/369 [03:11<00:00, 2.15it/s]
100%	737/737 [00:01<00:00, 375.65it/s]

Figure 10-5. *Output*

```
#Generating summaries on news test data
results = model.eval_model(test_df)

#print the loss
results

{'eval_loss': 2.0229070737797725}
```

There is a loss of 2.02. The model is trained. Now let's look at some of the results to evaluate the performance. Let's print both the original summary as well as the machine-generated summary.

```
#Original test data text summary for top 10 news

for i in test_df.target_text[:10]:
  print(i)
Astronaut reveals he once mistakenly called 911 from space
Try babysitting to survive govt shutdown: US Coast Guard to staff
Tejashwi touches brother Tej's feet amid rumours of a rift
Govt files FIR against NGO that organises Bravery Awards
Man sues Avengers actress Gwyneth for â?¹22 crore over ski crash
Hope India will seek new waiver from US to buy our oil: Iran
Apne dum pe khela hoon jitna khela hoon: Irfan Pathan to troll
Ex-IAF officer's son gets 10yrs in Saudi jail over tweet on Prophet
NZ women cricketers earn less than commentators in T20 tournament
Plane carrying Argentine footballer joining new team goes missing

#Predicted summary
for i in test_df.input_text:
  print(model.predict([i]))
```

Figure 10-6 shows the model training.

```
Generating outputs: 100% |████████████████████████| 1/1 [00:00<00:00, 4.34it/s]
["I once dialed 911 from space, didn't know it was international: Astronaut"]
Generating outputs: 100% |████████████████████████| 1/1 [00:00<00:00, 4.94it/s]
['US Coast Guard asks employees to consider babysitting, dog-walking amid shutdown']
Generating outputs: 100% |████████████████████████| 1/1 [00:00<00:00, 4.14it/s]
["Lalu Prasad's son Tej Pratap shares pics of meeting his"]
Generating outputs: 100% |████████████████████████| 1/1 [00:00<00:00, 5.81it/s]
['FIR against NGO running Bravery Awards for misusing funds']
Generating outputs: 100% |████████████████████████| 1/1 [00:00<00:00, 4.45it/s]
['Man sues Avengers actress Gwyneth Paltrow for crashing into him while skiing']
```

Figure 10-6. *Output*

We successfully trained the model using transfer learning. We used BART here. Similarly, let's use another T5 model to see what the output looks like.

T5 Pretrained Model

T5 is short for *text-to-text transfer transformer.* It's a transformer model from Google where it takes text as input and modified text as output. It is easy to fine-tune this model on any text-to-text task, which makes it very special.

This transformer model is trained on the Colossal Clean Common Crawl (C4) data set. The architecture is integrated to devise text2text model. And hence T5 can also be used for many text-to-text-based applications.

Figure 10-6 demonstrates that the task name must be in front of the input sequences, which goes to the T5 model, and the output is text.

Now, let's look at how to use Hugging Face's transformers library and implement news heading generation using the article information as input.

Let's install T5 first by running this command.

```
#Installing simple-T5
! pip install simplet5 -q

#Import the library
from simplet5 import SimpleT5
```

Let's prepare the data in a way the model expects.

```
# Model expects dataframe to have 2 column names
# Input as "source_text", Summary as "target_text", let us rename accordingly

df = df.rename(columns={"headlines":"target_text", " input_text":"source_
text"})
df = df[['source_text', 'target_text']]

# Let us add a prefix "summarize: " for all source text

df['source_text'] = "summarize: " + df['source_text']
df
```

	source_text	target_text
0	summarize: Saurav Kant, an alumnus of upGrad and IIIT-B's PG Program in Machine learning and Artificial Intelligence, was a Sr Systems Engineer at Infosys with almost 5 years of work experience. T...	upGrad learner switches to career in ML and AI with 90% salary hike
1	summarize: Kunal Shah's credit card bill payment platform, CRED, gave users a chance to win free food from Swiggy for one year. Pranav Kaushik, a Delhi techie, bagged this reward after spending 20...	Delhi techie wins free food from Swiggy for one year on CRED
2	summarize: New Zealand defeated India by 8 wickets in the fourth ODI at Hamilton on Thursday to win their first match of the five-match ODI series. India lost an international match under Rohit Sh...	New Zealand end Rohit Sharma-led India's 12-match winning streak
3	summarize: With Aegon Life iTerm Insurance plan, customers can enjoy tax benefits on your premiums paid and save up to â?¹46,800^ on taxes. The plan provides life cover up to the age of 100 years....	Aegon life iTerm insurance plan helps customers save tax
4	summarize: Speaking about the sexual harassment allegations against Rajkumar Hirani, Sonam Kapoor said, "I've known Hirani for many years...What if it's not true, the [#MeToo] movement will get de...	Have known Hirani for yrs, what if MeToo claims are not true: Sonam

```
#Splitting data into train and test

from sklearn.model_selection import train_test_split

train_df, test_df = train_test_split(df, test_size=0.2)
train_df.shape, test_df.shape

#Initializing the model
model = SimpleT5()

#Importing pretrained t5 model
model.from_pretrained(model_type="t5", model_name="t5-base")

#Import torch as this model is built on top of pytorch
import torch
torch.cuda.empty_cache()
```

Now that we have imported all the necessary libraries, let's train the model.

```
#Training the model with 5 epochs

model.train(train_df=train_df,
            eval_df=test_df,
            source_max_token_len=128,
            target_max_token_len=50,
            batch_size=8, max_epochs=5, use_gpu=True)

#Models built at each epoch
! ( cd outputs; ls )

SimpleT5-epoch-0-train-loss-1.5703   SimpleT5-epoch-3-train-loss-0.7293
SimpleT5-epoch-1-train-loss-1.1348   SimpleT5-epoch-4-train-loss-0.5978
SimpleT5-epoch-2-train-loss-0.902

#Loading model saved with lowest loss
model.load_model("t5","outputs/SimpleT5-epoch-4-train-loss-0.902",
use_gpu=True)
```

The model is now trained. Let's look at some predictions.

```
#Original headlines

test_df['target_text']

#output

2249      Batsman gets out on 7th ball of over in BBL, umpires don't notice
62       Woman rejects job after CEO bullies her in interview; he apologises
1791       SC allows dance bars to reopen in Mumbai; bans showering of cash
2341             PM is in panic: Ex-Uttarakhand CM on Alok Verma's removal
91        Whoever invented marriage was creepy as hell: Sushmita posts joke
                                  ...
294                Honor View20 unveiled, to go on sale in India from Jan 30
3319       India enforce follow-on, Aus trail by 316 runs after rain-hit day
813        Bharat needs Eid release, couldn't fit in time for ABCD 3: Katrina
1974          GST Council forms panel to resolve real estate sector issues
2891      Possibility that Vinta accused Alok of rape for own benefit: Court
```

```
#Model summarized headlines
for doc in test_df['source_text']:
  print(model.predict(doc))
['Klinger dismissed on 7th legal delivery after miscount by umpires']
['I was bullied to the point of tears: 22-yr-old writer turns down job
offer']
['SC paved way for dance bars in Mumbai']
['PM fearful, in panic: Harish after removal of CBI chief']
['Whoever invented marriage was creepy as hell: Sushmita Sen']
["Don't want to talk about Hirani sexual harassment row: Nawazuddin"]
['One alleged smuggler carrying 1,44,000 drugs held in Manipur']
```

Results are not bad, right? Now let's predict on test data as well some random sentences from the Internet.

```
#Generating random news summary on text from the internet
```

```
model.predict('summarize: As many as 32 died following a landslide in
Taliye village in Mahad Taluka, Raigad on Thursday. Around 35 houses were
buried under debris and several people were still feared to be missing or
trapped under it.  In another landslide, four people died in Poladpur,
which is also a landslide-prone area.')
```

```
['32 dead in landslide in Raigad village; 35 houses buried under debris']
#Checking headlines for test data
for doc in test_df['source_text']:
  print(model.predict(doc))
  print(())
```

['Punjab minister upgraded to Z plus, gets bullet-proof Land Cruiser']
summarize: Punjab minister Navjot Singh Sidhu's security cover has been
upgraded to Z plus and a bullet-proof Land Cruiser from CM Captain
Amarinder Singh's fleet has been given to him. Punjab government also asked
Centre to provide Sidhu with Central Armed Police Forces cover, claiming
that the "threat perception" to Sidhu increased after he attended Pakistan
PM Imran Khan's swearing-in ceremony.
['Asthana appointed chief of Bureau of Civil Aviation Security']
summarize: After the Appointments Committee of the Cabinet curtailed his
tenure as the CBI Special Director, Rakesh Asthana has been appointed as
the chief of Bureau of Civil Aviation Security. Asthana and three other
CBI officers' tenure was cut short with immediate effect. Last week, CBI
Director Alok Verma was moved to fire services department, following which
Verma announced his resignation.
['Marriage of Muslim man with Hindu woman merely irregular marriage: SC']
summarize: Citing Muslim law, the Supreme Court has said the marriage of a
Muslim man with a Hindu woman "is neither a valid nor a void marriage, but
is merely an irregular marriage". "Any child born out of such wedlock is
entitled to claim a share in his father's property," the bench said. The
court was hearing a property dispute case.
["I'm uncomfortable doing love-making scenes: Amrita Rao"]
summarize: Amrita Rao has said she's uncomfortable doing love-making
scenes, while adding, "Love-making is so personal to me...if I do it on
screen, it's like I'm leaving a part of my soul. I cannot do that," she

added, "I'm not saying it's wrong...it's the reflection of how our society has changed." "It's just about a choice...we all make," Amrita further said.

['11-hr bandh across northeastern states on Jan 8 against Citizenship bill']

summarize: Student organisations and indigenous groups have called for an 11-hour bandh across the northeastern states on January 8 from 5 am-4 pm in protest against Centre's decision to table Citizenship (Amendment) Bill, 2016 in Parliament. The bill seeks to grant citizenship to minority communities, namely, Hindus, Sikhs, Christians, Parsis, Buddhists and Jains from Bangladesh, Pakistan and Afghanistan.

['Ex-Gujarat CM joins Nationalist Congress Party']

summarize: Former Gujarat Chief Minister Shankersinh Vaghela joined the Nationalist Congress Party (NCP) on Tuesday in presence of party chief Sharad Pawar. The 78-year-old who left Congress in 2017, on Friday said, "In public life, a good platform is required to raise issues." "Vaghela... is a dynamic leader who knows the pulse of...the country," Gujarat NCP chief Jayant Patel had said.

The results look promising. We can further improve and polish the summary to handle noise in the output. First, let's evaluate the summarization model.

Evaluation Metrics for Summarization

- **ROUGE score**: Basically, the number of words in the human-generated summary appeared in the machine extracted summary. It measures the recall of the output in terms of machine learning evaluation.

- **BLEU score**: The score is calculated using the following formula.

  Score = number of words in the *machine-generated summary* that appeared in the human-created summary.

  The BLUE score gives the precision of the model.

Let's quickly find the BLUE score for the T5 model.
Let's import the library.

```
from nltk.translate.bleu_score import sentence_bleu
```

Let's predict for the source text, which is new.

```
x= [x for x in df.headlines]
y=[model.predict(p)[0] for p in df['source_text']]

# Function to calculate the score
L=0
for i,j in zip(x,y):
  L+=sentence_bleu(
    [i],
    j,
    weights=(0.25, 0.25, 0.25, 0.25),
    smoothing_function=None,
    auto_reweigh=False,
)

#Average blue score of whole corpuses
L/df.shape[0]
```

```
0.769163055697453
```

The average score for the whole corpus is 77%.

Future Scope

The output must be grammatically correct, readable, and well-formatted. For this to happen even after the summary is generated, we need to do output processing. This post-processing is done at the word or sentence level to make the output understandable.

The following are the few post-processing steps we can add to the pipeline.

1. Check grammar. Given the output is machine-generated, we can expect a lot of grammatical mistakes. We need to clean them up before showing them to the user.

   We can use libraries like GECToR (Grammatical Error Correction: Tag, Not Rewrite) by Grammarly (`https://github.com/grammarly/gector`).

2. Remove duplicates. Another biggest challenge in summarization is duplicating—words and sentences. We need to remove them as well.

3. Remove incomplete sentences. We need to filter out incomplete sentences generated by the model, if any.

Conclusion

With the ever-increasing content and news from all types of media and social media, text summarization is one of the most widely used applications of NLP and deep learning. We explored how to build that pipeline right from importing raw data, cleaning it, using the pretrained model for custom training, evaluating, and finally predicting (generating summaries). There is a lot of scope to add more data, change the parameters, and use some high-end GPUs for better training, which would yield great results.

Applications are broad in every industry, and we can already see the traction through several research projects going on in this area. The use of reinforcement learning coupled with deep learning promises better results, and research continues in that direction. There are a lot of APIs in place to summarize the text, but they are not that accurate and reliable. Text summarization will soon reach a whole new level with all these advances.

CHAPTER 11

Text Generation: Next Word Prediction

This chapter explores ways to generate text or predict the next word, given the sequence of previous words. Use cases or applications include word/sentence suggestions while typing an e-mail in Gmail or text messages in LinkedIn, and machines writing poems, articles, blogs, chapters of novels, or journal papers.

Figure 11-1 shows an e-mail to Adarsha. As the word *Hope* is typed, Gmail recommends the next set of words, *you are doing well*. This is the best application of text generation.

Figure 11-1. *Text generation example*

Text generation models are built using state-of-the-art deep learning algorithms, especially the variants of RNNs and attention networks.

Text generation is a seq2seq modeling problem. These generation models can be built at the character level, word level, grams level, sentence level, of paragraph level.

© Akshay Kulkarni, Adarsha Shivananda and Anoosh Kulkarni 2022
A. Kulkarni et al., *Natural Language Processing Projects*, https://doi.org/10.1007/978-1-4842-7386-9_11

This project explores creating a language model that generates text by implementing and forming state-of-the-art recurrent networks variants and pretrained models.

Problem Statement

Explore how to build a sequence-to-sequence model for text generation applications like next word prediction or a sentence prediction based on the previous word or set of words.

Approach: Understanding Language Modeling

seq2seq models are key elements of many natural language processing models, such as translation, text summarization, and text generation.

Here the model is trained to predict the next word or sequence of words.

Let's look at how to formulate a seq2seq model for text generation tasks like next word prediction. Let's assume that this is the data on which are training the model. Ideally, the source data is huge with all the possible sentences and combinations of words.

The following is an input example:

I use a MacBook for work, and I like Apple.

Table 11-1 shows how to transform this data into the required format for the next-word prediction model.

Table 11-1. *Data Transformation*

Input	Output (Target)
I	use
Use	MacBook
MacBook	for
For	my
My	work
Work	and
And	I
I	like
Like	Apple
I use	MacBook
use MacBook	for
….	
I use	MacBook for
…..	

We are creating input and output for all combinations of unigrams, bigrams, and so forth, sequentially. It can be one to many, many to one, or many to many to represent input and output accordingly; hence, it is called a seq2seq model.

Since capturing context is vital, the model that comes to mind is RNN and its variant. So, we can start with basic LSTM architecture and then explore architectures like bidirectional LSTM, autoencoders, attention mechanisms, and transformers.

Also, let's look at some of the latest state-of-the-art pretrained models for seq2seq and text generation models.

Here are the three logical flow diagrams for different ways of training seq2seq or text generation models.

- Training the model from scratch (see Figure 11-2)

- Transfer learning (see Figure 11-3)

- Directly using pretrained model (see Figure 11-4)

TRAINING FROM SCRATCH

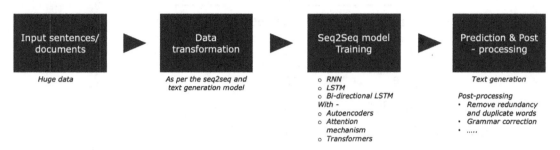

Figure 11-2. *Training from scratch*

TRANSFER LEARNING

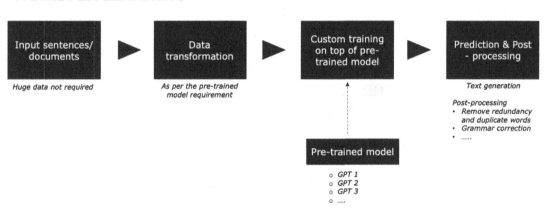

Figure 11-3. *Transfer learning*

DIRECTLY USING PRE-TRAINED MODEL

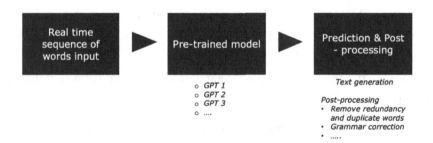

Figure 11-4. *Using pretrained model*

Now that you understand how to approach text generation models. Let's jump into different ways of implementation.

Implementation

First, let's explore few different approaches for developing a text-based language model from scratch using deep learning.

Model 1: Word-to-Word Text Generation

Models are trained to capture the relationship between the sequence of two words, considering the first word as input and the second one as target(output). Note: Import all the required libraries.

The first step is to convert the input sequence into a vector for each lowercase text in the source text.

```
# data for training (we picked few random paragraphs from the internet (blog)

#data source link 1: maggiesmetawatershed.blogspot.com
#data source link 2: www.humbolt1.com

text = """"She's a copperheaded waitress,
tired and sharp-worded, she hides
her bad brown tooth behind a wicked
smile, and flicks her ass
out of habit, to fend off the pass
that passes for affection.
She keeps her mind the way men
keep a knife—keen to strip the game
down. The ladies men admire, I've heard,
Would shudder at a wicked word.
Their candle gives a single light;
They'd rather stay at home at night.
They do not keep awake till three,
Nor read erotic poetry.
They never sanction the impure,
```

```
Nor recognize an overture.
They shrink from"""
```

```
# text encoding
# To tokenize the sentence let's use Tokenizer function
token = Tokenizer()
token.fit_on_texts([text])
#convert text to features
ohe = token.texts_to_sequences([text])[0]
```

Let's build the vocabulary dimension.

```
# text vocab dim
text_dim = len(token.word_index) + 1
```

Then we need to create a training data with input and output words.

```
# word to word sequence
#create empty list
seq = list()
for k in range(1, len(ohe)):
    seq1 = ohe[k-1:k+1]
#append to list
    seq.append(seq1)
```

```
# convert text into input and output
seq = np.array(seq)
I, O = seq[:,0],seq[:,1]
```

Let's encode the output.

```
# encoding the y variable
O = keras.utils.np_utils.to_categorical(O, num_classes=text_dim)
```

Now let's define a seq2seq model.

```
#initialize
mod = Sequential()
#add embedding layer
mod.add(Embedding(text_dim, 10, input_length=1))
```

```
#add LSTM layer
mod.add(LSTM(50))
#add dense layer
mod.add(Dense(text_dim, activation='softmax'))
print(mod.summary())
```

Figure 11-5 shows the model training output.

Layer (type)	Output Shape	Param #
embedding_2 (Embedding)	(None, 1, 10)	770
lstm_2 (LSTM)	(None, 50)	12200
dense_2 (Dense)	(None, 77)	3927

```
Total params: 16,897
Trainable params: 16,897
Non-trainable params: 0
```

Figure 11-5. *Output*

Technically, we simulated a multi-class classification problem (predicting words in the vocabulary) so let's use a classification function with cross-entropy loss. Here we used Adam optimizer to optimize the cost function.

```
# run network
mod.compile(loss='categorical_crossentropy', optimizer='adam',
metrics=['accuracy'])
#Training
mod.fit(I, O, epochs=600)
```

```
#Output
```

```
Epoch 1/500
 - 1s - loss: 4.3442 - acc: 0.0104
Epoch 2/500
 - 0s - loss: 4.3426 - acc: 0.0625
Epoch 3/500
 - 0s - loss: 4.3414 - acc: 0.0521
Epoch 4/500
 - 0s - loss: 4.3403 - acc: 0.0625
```

After tuning the model, we test it by passing the words in the vocabulary and predicting the next word. We get the entire output of the prediction.

```
#input for the prediction
inp = 'ladies'
#convert to features
dummy = token.texts_to_sequences([inp])[0]
dummy = np.array(dummy)
#predict
y1 = mod.predict_classes(dummy)
for word, index in token.word_index.items():
    if index == y1:
        print(word)

#output
men
```

Now let's predict sequence of words.

```
# predicting the sequence of words
def gen_seq(mod, token, seed, n):
    inp, op = seed, seed
    for _ in range(n):
        # convert text to features
        dummy = token.texts_to_sequences([inp])[0]
        dummy = np.array(dummy)
        # prediction
        y1 = mod.predict_classes(dummy)
        outp = ''
        for word, index in token.word_index.items():
            if index == y1:
                outp = word
                break
        # append
        inp += ' ' + outp
    return inp
```

Let's use some elements of input to get a reasonable sequence output.

```
print(gen_seq(mod, token, 'wicked', 1))

#output:
wicked word
```

As the word *wicked* is typed, the model predicts the next set of sequence words: *word their candle gives a.*

This is reasonable given the small training scale. If you train on larger data and GPU, accuracy increases.

Model 2: Sentence by Sentence

Another method is to divide the input into a sequence of words (sentence by sentence). This approach helps you understand the context better since we are using a sequence of words. In this case, it is done to avoid the prediction of words on several lines. If we only want to model and generate text lines, then these words may be suitable for the moment.

First, let's tokenize.

```
# sentence to sentence sequences
#create empty list
seq = list()
for tex in text.split('\n'):
#convert text to features
    dum = token.texts_to_sequences([tex])[0]
    for k in range(1, len(dum)):
        seq1 = dum[:k+1]
#append to list
        seq.append(seq1)
```

Once we have the sequence, let's pad it to make it consistent in fixed length.

```
# padding
length = max([len(s) for s in seq])
seq = pad_sequences(seq, maxlen=length, padding='pre')
```

Then let's divide the sequence into the input and target.

```
# divide
seq = np.array(seq)
I, O = seq[:,:-1],seq[:,-1]
# encoding the output variable
O = keras.utils.np_utils.to_categorical(O, num_classes=text_dim)
```

The model definition and initialization code remain the same as the previous method, but the input sequence is more than a single word. So, change input_length to len-1 and rerun the same code.

Figure 11-6 shows the model training output.

```
Layer (type)                   Output Shape          Param #
=================================================================
embedding_3 (Embedding)        (None, 7, 10)         770

lstm_3 (LSTM)                  (None, 50)            12200

dense_3 (Dense)                (None, 77)            3927
=================================================================
Total params: 16,897
Trainable params: 16,897
Non-trainable params: 0
```

```
None
Epoch 1/500
 - 1s - loss: 4.3442 - acc: 0.0125
Epoch 2/500
 - 0s - loss: 4.3413 - acc: 0.0250
Epoch 3/500
 - 0s - loss: 4.3385 - acc: 0.0625
Epoch 4/500
 - 0s - loss: 4.3364 - acc: 0.0750
Epoch 5/500
 - 0s - loss: 4.3337 - acc: 0.0875
Epoch 6/500
 - 0s - loss: 4.3307 - acc: 0.1125
.............
```

Figure 11-6. *Output*

Now, let's use this trained model for generating new sequences as before.

```
# predicting the next word
def gen_seq(mod, token, length, seed, n):
```

```
    inp = seed

    for _ in range(n_words):
        #convert text to features
        dum = token.texts_to_sequences([inp])[0]

        dum = pad_sequences([dum], maxlen=length, padding='pre')
        #predict
        y1 = mod.predict_classes(dum)

        outp = ''
        for word, index in token.word_index.items():
            if index == y1:
                outp = word
                break

        inp += ' ' + outp
    return inp

print(gen_seq(mod, token, length-1, 'she hides her bad', 4))
print(gen_seq(mod, token, length-1, 'hides', 4))

#Output
she hides her bad the way men i've
hides bad shudder at a
```

Model 3: Sequence of Input Words and Output Word

Let's use three input words to predict an output word, as follows.

```
# encoding 2 words - 1 word
#create empty list
seq = list()
for k in range(2, len(ohe)):
    seq1 = ohe[k-2:k+1]
        #append to list
    seq.append(seq1)]
```

Model initialization and building step would remain the same as the previous one.

Let's see what the prediction looks like.

```
# prediction

print(gen_seq(mod, token, len-1, 'candle gives', 5))
print(gen_seq(mod, token, len-1, 'keeps her', 3))
```

Output:

```
candle gives a a wicked light light
keeps her mind the way
```

If you notice the output or next word prediction, it is not great. It is predicting the same word in some of the cases. However, if we train this model/framework on large machines with huge data, we can expect good results. Also, post-processing of the output is required, as discussed in Chapter 10.

To overcome any concerns, we can leverage transfer learning concepts to use pretrained models as we did for text summarization.

Before that, let's explore some other, bit bigger data set to build a text generation model using the built frameworks.

The data set contains sonnets of Mr. Shakespeare.

Figure 11-7 shows the top recurring words throughout the data set:

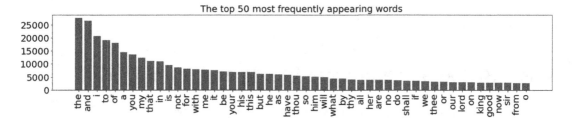

Figure 11-7. Output

Here, you can see that common words like *the, end,* and *i* were highly used in the input data. So, you can safely assume that the model is biased on these words.

Figure 11-8 shows the least occurring words in the data set.

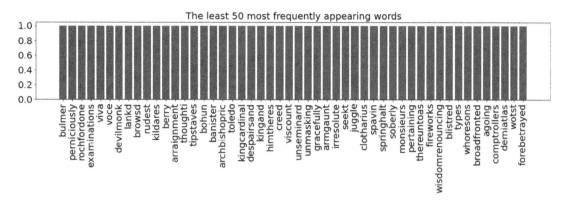

Figure 11-8. *Output*

Figure 11-8 shows the words that have been used very rarely in the data set, thus the model predicts words from these words very rarely.

As with previous frameworks, this includes tokenization of text, vocabulary and inverse vocabulary formation and building batches of sequences of input text.

```
##We use this function to clean text and tokenize it.
def clean_text(doc):
    toks = doc.split()
    tab = str.maketrans('', '', string.punctuation)
    toks = [words.translate(table) for words in toks]
    #consider he words which has only alphabets.
    toks =[words for words in toks if words.isalpha()]
    #Lower casing
    toks = [words.lower() for words in toks]
    return toks
```

Vocabulary is formed in which each word is mapped to an index based on sentences collections, The "counter" forms key/value pairs for each word of text. The "counter" is a container that stores words as keys and their count as values.

Inverse vocabulary is maintained in which there is an index to word mapping. It is done by using an enumerate function. This gives tuples in the form of counter and element.

We can leverage the same model definition code used in the previous section and change input length and vocab size according to this data set.

Figure 11-9 illustrates the workings of the model.

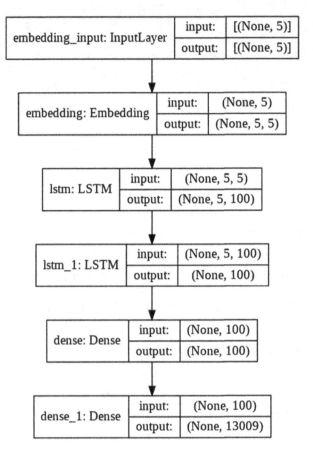

Figure 11-9. *Output*

After sampling, the new text is generated using the words from the input text, which is shown in the model. Since the model is trained on Shakespearean sonnets, the model can generate the upcoming words.

Figure 11-10 shows the output.

```
[ ] #Inputting seed_text --- required length 5 --- required words to be in vocabulary
    seed_text = lines[12343]
    print(seed_text)
    #Calling text generation
    print(seed_text + ' ' + generate_text_seq(model, tokenizer, seq_length, seed_text, 10)) #number of words to be predicted = 10

    home of love if i have
    home of love if i have ranged to leaving the witness green that longer i felt
```

```
    seed_text = lines[12300]
    print(seed_text)
    #Calling text generation
    print(seed_text + ' ' + generate_text_seq(model, tokenizer, seq_length, seed_text, 10)) #number of words to be predicted = 10

    outward form would show it dead
    outward form would show it dead and keep me out with wondrous that found wherein haste
```

```
[ ] seed_text = lines[5400]
    print(seed_text)
    #Calling text generation
    print(seed_text + ' ' + generate_text_seq(model, tokenizer, seq_length, seed_text, 10)) #number of words to be predicted = 10

    use it might unused stay from
    use it might unused stay from the courtiers breed as speech as messengers is corn and
```

```
[ ] seed_text = lines[48]
    print(seed_text)
    #Calling text generation
    print(seed_text + ' ' + generate_text_seq(model, tokenizer, seq_length, seed_text, 10)) #number of words to be predicted = 10

    lies thy self thy foe to
    lies thy self thy foe to every better name and swear he lends upon the earth
```

```
[ ] #Inputting seed_text --- required length 5 --- required words to be in vocabulary
    seed_text = lines[7272]
    print(seed_text)
    #Calling text generation
    print(seed_text + ' ' + generate_text_seq(model, tokenizer, seq_length, seed_text, 10)) #number of words to be predicted = 10

    time will come and take my
    time will come and take my love with his preferment clutchd me days he would not
```

Figure 11-10. *Output*

So far, you have seen to train a seq2seq model for text generation tasks. We can leverage the same frameworks to train on huge data and GPU for better outcomes and add more layers to the model architecture like bidirectional LSTM, dropouts, encoder-decoder, and attention.

Next, let's look at how to leverage some of the state-of-the-art pretrained models for text generation tasks.

GPT-2 (Advanced Pretrained Model)

GPT-2 stands for Generative Pretrained Transformer version-2. It is a huge architecture built on transformers with billions of hyperparameters and trained with 50 GB of text on the Internet. The GPT-2 model is built to focus on the task of text generation. GPT-2 is an improved version of GPT-1. GPT outperforms other domain-specific models on paper.

For more information on GPT and its versions, refer to our book *Natural Language Processing Recipes: Unlocking Text Data with Machine Learning and Deep Learning Using Python* (Apress, 2021).

Explanation of GPT in detail is beyond the scope of this book. For more information on the GPT-1,2,3, please refer to the research papers from `https://openai.com/research/`.

There are four different versions of GPT-2 available.

This chapter uses a large model with approximately 700 million parameters. The transformers library for this model contains weights of different versions of it. This is a state-of-the-art model, and we want to use it as generalized as possible, so we won't be custom training here. We use it and make a customized function that gives output as per the given input words.

Import and initialize gpt2

```
!pip install transformers
```

```
import torch
import transformers
```

```
#import tokenization and pre trained model
ftrs = transformers.GPT2Tokenizer.from_pretrained('gpt2')
pre_g_model = transformers.GPT2LMHeadModel.from_pretrained('gpt2')
```

```
#input text
input_text="We are writing book on NLP, let us"
```

```
# tokenization for input text
input_tokens = ftrs.encode(input_text,return_tensors="pt")
```

```
#sequences generated from the model
seq_gen = pre_g_model.generate(input_ids=input_tokens,max_
length=100,repetition_penalty=1.4)
seq_gen = seq_gen.squeeze_()
```

```
# predicted sequences
seq_gen
```

```
tensor([ 1135,    389,   3597,   1492,    319,    399, 19930,     11,   1309,    514,
          760,    644,    345,    892,     13,    198,    464,    717,   1517,    356,
          761,    284,    466,    318,    651,    262,   1573,    503,    546,    340,
          290,    787,   1654,    326,    661,   1833,    703,   1593,    428,   2071,
         1107,    373,    329,    606,    526, 50256])
```

```
# lets get the words from decoder

txt_gen = []

for i, j in enumerate(seq_gen):
        j = j.tolist()
        txt = ftrs.decode(j)
        txt_gen.append(txt)

# final text generated

text_generated = ''
text_generated.join(txt_gen).strip('\n')
#output
```

We are writing a book on NLP; let us know what you think.

The first thing we need to do is get the word out about it and make sure that people understand how important this issue is to them."<|endoftext|>

The example gives meaningful output. We haven't fine-tuned the model or re-trained on any data, and it derives context and predicts the next set of words. Here you can change parameters according to your needs where max_length refers to the maximum number of words to predict, and n_seqs refers to the number of different sequences (i.e., the number of different outputs from the same input, as required). The results we got were quite interesting.

So, to sum it up, this model outperformed our expectations for the text generation task.

Note   The architecture of GPT-3 is similar to GPT-2. Few major differences are around the attention layers, word embedding size, and activation function parameters.

Now, let's move on to another interesting library that can be leveraged in most applications that requires text input. This intelligent system saves time by recommending what to write in a search engine or a WhatsApp message, for example.

Autocomplete/Suggestion

We already worked on few text generation models in which few were made locally trained on a relatively very small database. The other was GPT-2 state-of-the-art NLP generalized model to generate text word by word. Now it is time to predict the upcoming word from half-written/ill-written characters. Ultimately, we wanted to look for "Incremental Search AKA Auto Complete AKA Auto Suggestion". Here, as you type text, possible matches are found and immediately presented to you. Most of these applications are enabled in search engines.

Figure 11-11 shows how an autoword suggestion works.

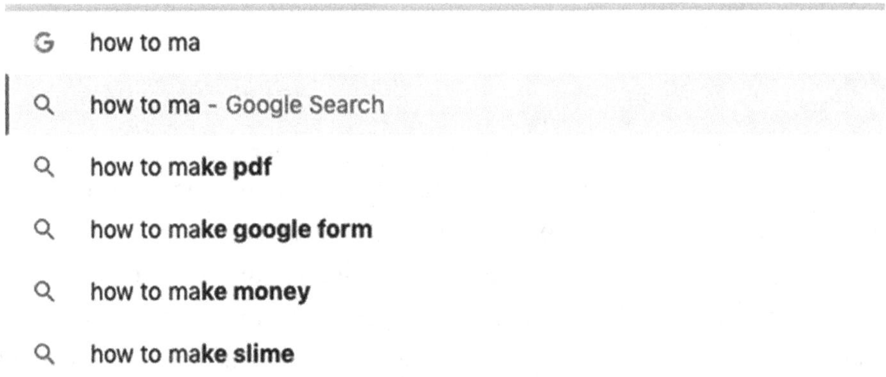

G   how to ma

Q   **how to ma** - Google Search

Q   how to ma**ke pdf**

Q   how to ma**ke google form**

Q   how to ma**ke money**

Q   how to ma**ke slime**

Figure 11-11. *Example of autosuggestion*

Fast Autocomplete

The Fast Autocomplete library (by zepworks) efficiently searches a corpus of words. This library is completely based on Python language. Internally, it makes a trie tree of every unique word in the corpus. The data structure they used in this library is called Dwg, which stands for *directed word graph*.

Install the library:

```
!pip install fast-autocomplete
```

To use this library, we needed to extract tokens from the data and put them in a specific format, which acts as the context to the model for providing suggestions as the user types.

Let's use the same data we used previously, which contained sonnets of Shakespeare, and write the function to extract words from it.

The following is the function to extract a unique token from the corpus.

```
def unique_tokens(location):
to_doc = read_csv_gen(location, csv_func=csv.DictReader)
     toks = {}
     for row in to_doc:
                distinct_words = row['distinct_words']
                index = row['index']
                if distinct_words != index:
                          local_words = [index, '{} {}'.format
                          (distinct_words, index)]
                          while local_words:
                                 w = local_words.pop()
                                 if w not in toks:
                                        toks[w] = {}
                if distinct_words not in toks:
                     toks[distinct_words] = {}
     return toks

print(toks)
```

Figure 11-12 shows tokens output.

```
{'missions': {}, 'garrisond': {}, 'doublehornd': {}, 'brotherhood': {},
◄ ▇
```

Figure 11-12. *Output*

Similarly, we can provide a predefined dictionary for synonyms, if any.

Finally, we initialize the autocomplete model.

```
from fast_autocomplete import AutoComplete
autocomplete = AutoComplete(words=toks)
```

Now that we have provided the data context to the model, we only need to call the search function from autocomplete library to get the auto word suggestion based on the context/data provided. It takes only three parameters.

- Word: the word you want to type

- Max_Cost: Distance metric cost

- Size: no. of words to recommend

Figure 11-13 shows searching a word like *straw*. The recommendations include *straws* and *strawberry*. Similarly, we experimented with *tr*, *hag*, *sc*, and *mam*.

```
[ ]   #Output of straw
      autocomplete.search(word='straw', max_cost=3, size=10)

      [['straw'],
       ['straws'],
       ['strawy'],
       ['strawberry'],
       ['strawcolour'],
       ['strawberries']]
```

```
●   #Output of tr
    autocomplete.search(word='tr', max_cost=3, size=10)
```

```
●   [['try'],
     ['trop'],
     ['troy'],
     ['trow'],
     ['trot'],
     ['trod'],
     ['trap'],
     ['tray'],
     ['tres'],
     ['tree']]
```

```
[ ]   #Output of hag
      autocomplete.search(word='hag', max_cost=3, size=10)

      [['hag'],
       ['hags'],
       ['hagars'],
       ['hagseed'],
       ['haggard'],
       ['haggled'],
       ['haggish'],
       ['haggards'],
       ['hagbornnot']]
```

```
[ ]   #Output of sc
      autocomplete.search(word='sc', max_cost=3, size=10)

      [['scum'],
       ['scut'],
       ['scot'],
       ['scan'],
       ['scar'],
       ['scab'],
       ['scuse'],
       ['scour'],
       ['scout'],
       ['scoff']]
```

```
[ ]   #Output of mam
      autocomplete.search(word='mam', max_cost=3, size=10)

      [['mammet'], ['mammets'], ['mammockd'], ['mammering'], ['mamillius']]
```

Figure 11-13. *Output*

You saw how single-word suggestion works. Similarly, you can implement multiple sequences of word suggestions. For more information, refer to `https://pypi.org/project/fast-autocomplete/`.

Conclusion

We implemented different ways to model language for text generation using different techniques and libraries. All these are the baseline models, can be improved with huge training data and bigger GPUs.

There is ample application of text generation for which we can utilize the frameworks discussed in this chapter and use domain-specific training data accordingly based on the use case. One amazing Git application generates code from the comments and the other way around.

You have seen and implemented few complex algorithms and learned some state of art algorithms in NLP and deep learning. Chapter 12 discusses what's next in NLP and deep learning. Where to go from here? We uncover some of the interesting research areas.

CHAPTER 12

Conclusion and Future Trends

You have learned how to build different NLP applications and projects leveraging the power of machine learning and deep learning, which helps solve business problems across industries.

This book focused on aspects of

- Classification (e-commerce product categorization, complaint classification, Quora duplication prediction)

- Clustering (TED Talks segmentation and other document clustering)

- Summarization (news headlines summarization)

- Topic extraction (résumé parsing and screening)

- Entity recognition (CRF extracting custom entitles from movie plot)

- Similarity (résumé shortlisting and raking)

- Information retrieval (AI-based semantic search, how smart search engines can improve the customer experience for e-commerce)

- Generation (next word prediction, autocomplete)

- Information extraction (résumé parsing)

- Translation (multilanguage for search)

- Conversational AI (chatbots and Q&A)

You also looked at how some of the latest SOTA algorithms build more robust models and applications.

Where to go from here?

© Akshay Kulkarni, Adarsha Shivananda and Anoosh Kulkarni 2022
A. Kulkarni et al., *Natural Language Processing Projects*, https://doi.org/10.1007/978-1-4842-7386-9_12

NLP and deep learning are very active in research. As you are reading this, many new algorithms are being designed to increase the intelligence of applications and make NLP more efficient and user-friendly.

More information on next-generation NLP can be found in our book *Natural Language Processing Recipes: Unlocking Text Data with Machine Learning and Deep Learning Using Python* (Apress, 2019).

The latest and future trends around NLP can be categorized into two aspects.

- Depth (tech and algorithms)

- Broad/wide (applications and use cases)

This book briefly looked at a few active research areas (depth aspect).

- Advancements in word embeddings. You looked at word2vec, GloVe, BERT, Transformers, and GPT. We can watch this section for more SOTA text to features algorithms improvised on top of transformers and reformers (most recent and advanced transformer variant with reversible layers).

- Advanced deep learning for NLP: Current state is around LSTM, GRU, Bi-directional LSTM, auto-encoders, attention mechanism, transformers, and the auto-regressive layer. In the coming years, we can expect more efficient and improved deep learning and transformers architecture designed for NLP tasks.

- Reinforce learning application in NLP.

- Transfer learning and pretrained models: We used many SOTA pretrained models to solve summarization, generation, and search problems. We can expect many more generalized, robust, efficient, and more accurate pretrained models and transfer learning frameworks in the near future.

- Meta-learning in NLP.

- Capsule networks for NLP: Known for multitasking capabilities, a single trained model can solve diversified problems and perform multiple NLP tasks.

- Integrating/coupling supervised and unsupervised methods: Training a model for any task requires huge data and the labeled one.

The biggest challenge in this era is to get accurately labeled data. It's usually a manual process, but doing it on huge data requires a lot of time and resources. Hence there is active research to combine unsupervised and supervised to solve the challenge of labeled data for the model training.

Besides these tech deep areas, there are few areas from a broader or wider NLP perspective.

- Automation in NLP: AutoNLP

- Text can be in any language: Multilingual NLP

- Conversational AI

- Domain and industry-specific trained models

- NLP coupled with computer vision

Let's look at each of them.

AutoNLP

As a data scientist, it takes a lot of skill and time to create a good machine learning and deep learning model to train and implement various NLP-related tasks. One can face challenges in selecting the right parameters, optimizing, debugging, and testing them on new data. This is not much of a problem for a skilled data scientist but is still time-consuming. What if we can use something that automates everything using a framework that provides visualizations and gives excellent results with minimal error?

Hugging Face's AutoNLP is the way to go!

What is AutoNLP?

It lets you preprocess, train, tune and evaluate different NLP models or algorithms for various tasks hassle-free.

Hugging Face is a next-gen NLP start-up helping professionals and companies build and experiment state of the art NLP and deep learning models with ease. There is a lot you can explore and research at `https://huggingface.co/`.

The following are the main features of AutoNLP.

- Automatic selection of best models given your data

- Automatic fine-tuning

- Automatic hyperparameter optimization

- Model comparison after training

- Immediate deployment after training

- CLI and Python API available

It supports binary classification, multiclass classification, entity extraction, text summarization, Q&A, and speech recognition.

Figure 12-1 shows the capability of Hugging Face AutoNLP with respect to the tasks it can perform and languages it can support

Task	Languages Supported
Binary Classification	English, French, German, Finnish, Hindi, Spanish, Chinese, Dutch, Italian, Japanese, Bengali, Arabic
Multi-class Classification	English, French, German, Finnish, Hindi, Spanish, Chinese, Dutch, Italian, Japanese, Bengali, Arabic
Entity Extraction	English, French, German, Finnish, Hindi, Spanish, Chinese, Dutch, Italian, Japanese, Bengali, Arabic
Summarization	English, French, German, Finnish, Hindi, Spanish, Chinese, Dutch, Italian, Japanese, Arabic (Bengali is not supported)
Extractive Question Answering	English, French, German, Finnish, Hindi, Spanish, Chinese, Dutch, Italian, Japanese, Bengali, Arabic
Speech Recognition	English, French, German, Finnish, Hindi, Spanish, Chinese, Dutch, Italian, Japanese, Bengali, Arabic and 100 more.

Figure 12-1. Hugging Face

So, it supports multiple languages, which is another trend.

AutoNLP can make our lives much easier by performing various NLP-related tasks with minimum human intervention. We are in an era where technology is developing day by day. Through NLP, we are making human-to-machine communication as personal as human-to-human communication. So, there are many NLP trends we can look forward to in the coming years.

In addition to AutoNLP, Hugging Face has multiple other custom SOTA pretrained models, frameworks, and API.

In AutoNLP, you can also fine-tune a model that is hosted on Hugging Face Hub. There are more than 14,000 models available, which can be filtered based on the task you want to perform, your preferred language, or the data set (if hosted on Hugging Face) you wish to use.

The Hugging Face data sets library currently has over 100 public data sets.

The following are some of the models from Hugging Face.

- BERT-base

- RoBERTa

- DistilBERT

- Sentence Transformers

- GPT-2

- T5-base

- ALBERT

There are around 14,000 (increases every day) models. And you can apply filters to get models as per your requirements.

The following are some of the Hugging Face summarization task models.

- Distilbart-xsum

- Bart-large-cnn

- Pegasus-large

- Mt5

Let's move to the next trending technique.

Multilingual NLP

The following are the most prominent multilingual models.

- mBERT

- XLM

- XLM-R

- Multilingual BERT (mBERT): This is released along with the BERT and it supports 104 languages. Essentially it is BERT trained in many languages.

- XLM and XLM-R: XLM-R trains RoBERTa on a huge multilingual data set at an enormous scale.

Most NLP advances have been in English until recently. Major tech organizations like Google and Facebook are rolling out preskilled multilingual systems whose overall performance is on par with English.

On Facebook, if a person posts something on his timeline that is written in Chinese, there is an option to translate it to English.

Initially, Amazon's Alexa understood only the English language. Now it is extended to a Hindi version for India.

So, these two examples use the multilingual NLP models to understand different languages.

Figure 12-2 shows some of the most recent models created in the field of multilingual NLP, with the top results according to Google XTREME.

Rank	Model	Participant	Affiliation	Attempt Date	Avg	Sentence-pair Classification	Structured Prediction	Question Answering	Sentence Retrieval
0		Human	-	-	93.3	95.1	97.0	87.8	-
1	Polyglot	MLNLC	ByteDance	Apr 29, 2021	81.7	88.3	80.6	71.9	90.8
2	Unicoder + ZCode	MSRA + Cognition	Microsoft	Apr 26, 2021	81.6	88.4	76.2	72.5	93.7
3	VECO	DAMO NLP Team	Alibaba	Mar 22, 2021	81.4	88.9	75.6	72.9	92.7
4	ERNIE-M	ERNIE Team	Baidu	Jan 1, 2021	80.9	87.9	75.6	72.3	91.9
5	HiCTL	DAMO MT Team	Alibaba	Mar 21, 2021	80.8	89.0	74.4	71.9	92.6
6	T-ULRv2 + StableTune	Turing	Microsoft	Oct 7, 2020	80.7	88.8	75.4	72.9	89.3
7	Anonymous3	Anonymous3	Anonymous3	Jan 3, 2021	79.9	88.2	74.6	71.7	89.0
8	FILTER	Dynamics 365 AI Research	Microsoft	Sep 8, 2020	77.0	87.5	71.9	68.5	84.4
9	X-STILTs	Phang et al.	New York University	Jun 17, 2020	73.5	83.9	69.4	67.2	76.5
	XLM-R	XTREME	Alphabet,						

Figure 12-2. *Multilingual NLP*

As of August 2021, Polyglot stands at the top of the Google XTREME leaderboard with the highest points compared to other multilingual models.

Conversational AI

A few years back, it was all the human intelligence behind multiple domains to get the job done, like a customer support center.

And now, there is an artificially enabled intelligent system chatbot that does many of the things humans do.

These days, Amazon Alexa, Apple Siri, and Google Home are in every other home. Be it for your daily news, weather updates, entertainment, or traffic, we almost interact with these intelligent systems. In time, their performance will only improve.

Yes, this is one such topic hot in the air right now.

Industry-Specific Pretrained Models

Industry or domain-specific models start getting a lot of traction. Although many generalized pretrained models, these sometimes fail to perform well on domain-specific problems (whose data are not easily available on the Internet at scale).

For example, 80% of all health care data is locked in unstructured data. Its applications include the following.

- Information extraction from clinical notes and reports

- NLP in patient trials

There are several pretrained AI models.

- Pretrained sentiment models

- Pretrained chatbot models

- Pretrained text generation or summarization models

These might not perform well in the health care domain. Hence there is a need for custom training and making it domain-specific. There is a pretrained model called bioBERT that is domain-specific language representation on medical data.

Image Captioning

When two states of the art technologies combine, imagine the wonders they can do. Computer vision and NLP form a deadly combination for many human-like intelligent systems.

One such application is image caption generation or image-to-text summarization. Likewise, video summarization can be implemented that summarizes the scene of the video.

Figure 12-3 shows example 1.

Figure 12-3. *Example (photo by XPS on Unsplash* `https://unsplash.com/license`*)*

Figure 12-4 shows example 2.

Figure 12-4. *Example (photo by Harley-Davidson on Unsplash* `https://unsplash.com/license`*)*

Figure 12-5 shows example 3.

Figure 12-5. *Example (photo by Digital Marketing Agency NTWRK on Unsplash* `https://unsplash.com/license)`

There will be many applications under this umbrella.

We hope you enjoyed this book and are now ready to leverage the learnings to solve real-world business problems using NLP.

See you soon in the next book/edition.

Index

© Akshay Kulkarni, Adarsha Shivananda and Anoosh Kulkarni 2022
A. Kulkarni et al., *Natural Language Processing Projects*, https://doi.org/10.1007/978-1-4842-7386-9

ated in the United States
Baker & Taylor Publisher Services